CAMBRIDGE LIBRARY COLLECTION

Books of enduring scholarly value

Earth Sciences

In the nineteenth century, geology emerged as a distinct academic
discipline. It pointed the way towards the theory of evolution, as scientists
including Gideon Mantell, Adam Sedgwick, Charles Lyell and Roderick
Murchison began to use the evidence of minerals, rock formations and
fossils to demonstrate that the earth was older by millions of years than the
conventional, Bible-based wisdom had supposed. They argued convincingly
that the climate, flora and fauna of the distant past could be deduced from
geological evidence. Volcanic activity, the formation of mountains, and the
action of glaciers and rivers, tides and ocean currents also became better
understood. This series includes landmark publications by pioneers of the
modern earth sciences, who advanced the scientific understanding of our
planet and the processes by which it is constantly re-shaped.

Earthquakes and Other Earth Movements

John Milne (1850–1913) was a professor of mining and geology at the
Imperial College of Engineering, Tokyo. While living in Japan, Milne
became very interested in seismology, prompted by a strong seismic
shock he experienced in Tokyo in 1880. Sixteen years later Milne and two
colleagues completed work on the first seismograph capable of recording
major earthquakes. This book, originally published in London in 1886,
explains why earthquakes happen and what effects they have on land and in
the oceans. As Milne points out, Japan provided him with 'the opportunity
of recording an earthquake every week'. Starting with an introduction
examining the relationship of seismology to the arts and sciences, the
book includes chapters on seismometry, earthquake motion, the causes of
earthquakes, and their relation to volcanic activity, providing a thorough
account of the state of knowledge about these phenomena towards the end of
the nineteenth century.

Cambridge University Press has long been a pioneer in the reissuing of out-of-print titles from its own backlist, producing digital reprints of books that are still sought after by scholars and students but could not be reprinted economically using traditional technology. The Cambridge Library Collection extends this activity to a wider range of books which are still of importance to researchers and professionals, either for the source material they contain, or as landmarks in the history of their academic discipline.

Drawing from the world-renowned collections in the Cambridge University Library, and guided by the advice of experts in each subject area, Cambridge University Press is using state-of-the-art scanning machines in its own Printing House to capture the content of each book selected for inclusion. The files are processed to give a consistently clear, crisp image, and the books finished to the high quality standard for which the Press is recognised around the world. The latest print-on-demand technology ensures that the books will remain available indefinitely, and that orders for single or multiple copies can quickly be supplied.

The Cambridge Library Collection will bring back to life books of enduring scholarly value (including out-of-copyright works originally issued by other publishers) across a wide range of disciplines in the humanities and social sciences and in science and technology.

Earthquakes and Other Earth Movements

John Milne

CAMBRIDGE
UNIVERSITY PRESS

CAMBRIDGE UNIVERSITY PRESS

Cambridge, New York, Melbourne, Madrid, Cape Town,
Singapore, São Paolo, Delhi, Tokyo, Mexico City

Published in the United States of America by Cambridge University Press, New York

www.cambridge.org
Information on this title: www.cambridge.org/9781108072632

This edition first published 1886
This digitally printed version 2011

ISBN 978-1-108-07263-2 Paperback

THE

INTERNATIONAL SCIENTIFIC SERIES

VOL. LVI.

EARTHQUAKES

AND

OTHER EARTH MOVEMENTS

BY

JOHN MILNE

PROFESSOR OF MINING AND GEOLOGY
IN THE IMPERIAL COLLEGE OF ENGINEERING, TOKIO, JAPAN

WITH THIRTY-EIGHT FIGURES

LONDON
KEGAN PAUL, TRENCH, & CO., 1 PATERNOSTER SQUARE
1886

PREFACE.

In the following pages it has been my object to give a systematic account of various Earth Movements.

These comprise *Earthquakes*, or the sudden violent movements of the ground; *Earth Tremors*, or minute movements which escape our attention by the smallness of their amplitude; *Earth Pulsations*, or movements which are overlooked on account of the length of their period; and lastly, *Earth Oscillations*, or movements of long period and large amplitude which attract so much attention from their geological importance.

It is difficult to separate these Earth Movements from each other, because they are phenomena which only differ in degree, and which are intimately associated in their occurrence and in their origin.

Because Earthquakes are phenomena which have attracted a universal attention since the earliest times, and about them so many observations have been made, they are treated of at considerable length.

As very much of what might be said about the other Earth Movements is common to what is said about Earthquakes, it has been possible to make the description of these phenomena comparatively short.

The scheme which has been adopted will be understood from the following table :—

I. EARTHQUAKES.

1. Introduction.
2. Seismometry.
3. Earthquake Motion . .
   - (a) Theoretically.
   - (b) As deduced from experiments.
   - (c) As deduced from actual Earthquakes.
4. Earthquake Effects . .
   - (a) On land.
   - (b) In the ocean.
5. Determination of Earthquake origins.
6. Distribution of Earthquakes.
   - (a) In space.
   - (b) In time (geological time, historical time, annual, seasonal, diurnal, &c.)
7. Cause of Earthquakes.
8. Earthquake prediction and warning.

II. EARTH TREMORS.
III. EARTH PULSATIONS.
IV. EARTH OSCILLATIONS.

In some instances the grouping of phenomena according to the above scheme may be found inaccurate, as, for example, in the chapters referring to the effects and causes of Earthquakes.

This arises from the fact that the relationship between Earthquakes and other Earth phenomena are not well understood. Thus the sudden elevation of a coast line and an accompanying earthquake may be related, either as effect and cause, or *vice versâ*, or they may both be the effect of a third phenomenon.

Much of what is said respecting Earthquake motion will show how little accurate knowledge we have about these disturbances. Had I been writing in England, and, therefore, been in a position to make references to libraries and persons who are authorities on subjects connected with Seismology, the following pages might have been made more complete, and inaccuracies avoided. A large proportion of the material embodied in the following pages is founded on experiments and observations made during an eight years' residence in Japan, where I have had the opportunity of recording an earthquake every week.

The writer to whom I am chiefly indebted is Mr. Robert Mallet. Not being in a position to refer to original memoirs, I have drawn many illustrations from the works of Professor Karl Fuchs and M. S. di Rossi. These, and other writers to whom reference has been made, are given in an appendix.

For seeing these pages through the press, my thanks are due to Mr. Thomas Gray, who, when residing in Japan, did so much for the advancement of observational Seismology.

For advice and assistance in devising experiments, I tender my thanks to my colleagues, Professor T. Alexander, Mr. T. Fujioka, and to my late colleague, Professor John Perry.

For assistance in the actual observation of Earthquakes, I have to thank my friends in various parts of Japan, especially Mr. J. Bissett and Mr. T. Talbot, of Yokohama. For assistance in obtaining information from

Italian sources I have to thank Dr. F. Du Bois, from German sources Professor C. Netto, and from Japanese sources Mr. B. H. Chamberlain. For help in carrying out experiments, I am indebted to the liberality of the British Association, the Geological Society of London, the Meteorological and Telegraph departments of Japan, and to the officers of my own institution, the Imperial College of Engineering.

And, lastly, I offer my sincere thanks to those gentlemen who have taken part in the establishment and working of the Seismological Society of Japan, and to my publishers, whose liberality has enabled me to place the labours of residents in the Far East before the European public.

JOHN MILNE.

TOKIO, JAPAN : *June* 30, 1883.

CONTENTS.

CHAPTER I.

INTRODUCTION.

CHAPTER II.

SEISMOMETRY.

CHAPTER III.

EARTHQUAKE MOTION DISCUSSED THEORETICALLY.

CHAPTER IV.

EARTHQUAKE MOTION AS DEDUCED FROM EXPERIMENT.

CHAPTER V.

EARTHQUAKE MOTION AS DEDUCED FROM OBSERVATION
ON EARTHQUAKES.

CHAPTER XIII.

CHAPTER XIV.

DISTRIBUTION OF EARTHQUAKES IN TIME (*continued*).

CHAPTER XV.

BAROMETRICAL FLUCTUATIONS AND EARTHQUAKES—FLUCTUATIONS IN TEMPERATURE AND EARTHQUAKES

CHAPTER XVI.

RELATION OF SEISMIC TO VOLCANIC PHENOMENA.

CHAPTER XVII.

THE CAUSE OF EARTHQUAKES.

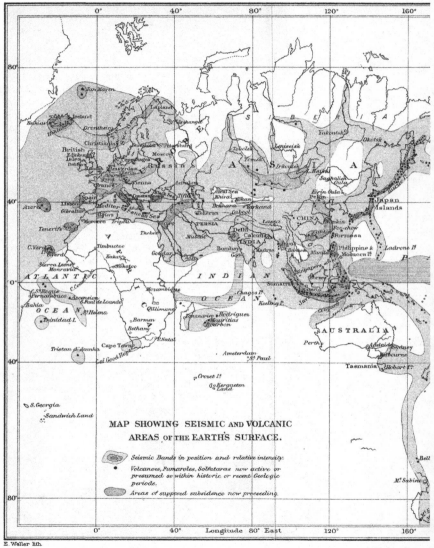

MAP SHOWING SEISMIC AND VOLCANIC
AREAS OF THE EARTH'S SURFACE.

Seismic Bands in position and relative intensity.

Volcanoes, Fumaroles, Solfataras now active or
presumed so within historic or recent Geologic
periods.

Areas of supposed subsidence now proceeding.

E. Weller lith.

London; Kegan P

Paul, Trench & Co.

EARTHQUAKES.

CHAPTER I.

INTRODUCTION.

Relationship of man to nature—The aspect of a country is dependent on geological phenomena—Earthquakes an important geological phenomenon—Relationship of seismology to the sciences and arts—Earth movements other than earthquakes—Seismological literature —(Writings of Perrey, Mallet, Eastern writings, the Philosophical Transactions of the Royal Society, the 'Gentleman's Magazine,' the Bible, Herodotus, Pliny, Hopkins, Von Hoff, Humboldt, Schmidt, Seebach, Lasaulx, Fuchs, Palmieri, Bertelli, Seismological Society of Japan)—Seismological terminology.

IN bygone superstitious times lightning and thunder were regarded as supernatural visitations. But as these phenomena became better understood, and men learned how to avoid their destructive power, the superstition was gradually dispelled. Thus it is with Earthquakes: the more clearly they are understood, the more confident in the universality of law will man become, and the more will his mental condition be advanced.

In his 'History of Civilisation in England,' Buckle has laid considerable stress upon the manner in which earthquakes, volcanoes, and other of the more terrible forms in which the workings of nature reveal themselves

B

to us, have exerted an influence upon the imagination and understanding; and just as a sudden fright may affect the nerves of a child for the remainder of its life, we have in the annals of seismology records which indicate that earthquakes have not been without a serious influence upon the mental condition of whole communities.

To a geologist there are perhaps no phenomena in nature more interesting than earthquakes, the study of which is called Seismology. Coming, as shocks often will, from a region of volcanoes, the study of these disturbances may enable us to understand something about the nature and working of a volcano. As an earthquake-wave travels along from strata to strata, if we study its reflections and changing velocity in transit, we may often be led to the discovery of certain rocky structures buried deep beneath our view, about which, without the help of such waves, it would be hopeless ever to attain any knowledge.

By studying the propagation of earthquake-waves the physicist is enabled to confirm his speculations respecting the transmission of disturbances in elastic media. For the physicist earthquakes are gigantic experiments which tell him the elastic moduli of rocks as they exist in nature, and when properly interpreted may lead him to the proper comprehension of many ill-understood phenomena. It is not impossible that seismological investigation may teach us something about the earth's magnetism, and the connection between earthquakes and the ' earth currents ' which appear in our telegraph wires. These and numerous other kindred problems fall within the domain of the physicist.

It is of interest to the meteorologist to know the connections which probably exist between earthquakes and the fluctuations of the barometer, the changes of the thermometer, the quantity of rainfall, and like phenomena to which he devotes his attention.

Next we may turn to the more practical aims of seismology and ask ourselves what are the effects of earthquakes upon buildings, and how, in earthquake-shaken countries, the buildings are to be made to withstand them. Here we are face to face with problems which demand the attention of engineers and builders. To attain what we desire, observation, common sense, and subtle reasoning must be brought to bear upon this subject.

In the investigation of the principle on which earthquake instruments make their records, in the analysis of the results they give, in problems connected with astronomy, with physics, and with construction, seismology offers to the mathematician new fields for investigation.

A study of the effects which earthquakes produce on the lower animals will not fail to interest the student of natural history.

A study like seismology, which leads us to a more complete knowledge of earth-heat and its workings, is to be regarded as one of the corner-stones of geology. The science of seismology invites the co-operation of workers and thinkers in almost every department of natural science.

We have already referred to the influence exerted by earthquakes over the human mind. How to predict earthquakes, and how to escape from their dangers, are problems which, if they can be solved, are of extreme interest to the world at large.

In addition to the sudden and violent movements which we call earthquakes, the seismologist has to investigate the smaller motions which we call earth-tremors. From observations which have been made of late years, it would appear that the ground on which we dwell is incessantly in a state of tremulous motion.

A further subject of investigation which is before the seismologist is the experimental verification of the existence of what may be called 'earth-pulsations.' These are motions which mathematical physicists affirmed the existence of, but which, in consequence of the slowness of their period, have hitherto escaped observation.

The oscillations, or slow changes in the relative positions of land and sea, might also be included ; but this has already been taken up as a separate branch of geology.

These four classes of movements are no doubt interdependent, and seismology in the widest sense might conveniently be employed to include them all. In succeeding chapters we will endeavour to indicate how far the first three of these branches have been prosecuted, and to point out that which remains to be accomplished. It is difficult, however, to form a just estimate of the amount of seismological work which has been done, in consequence of the scattered and uncertain nature of many of the records. Seismology, as a science, originated late, chiefly owing to the facts that centres of civilisation are seldom in the most disturbed regions, and that earthquake-shaken countries are widely separated from each other.

As every portion of the habitable globe appears to have been shaken more or less by earthquakes, and as these phenomena are so terrible in their nature, we can readily understand why seismological literature is extensive. In the annals of almost every country which has a written history, references are made to seismic disturbances.

An idea of the attention which earthquakes have received may be gathered from the fact that Professor Alexis Perrey, of Dijon, who has published some sixty memoirs on this subject, gave, in 1856, a catalogue of 1,837 works devoted to seismology.[1] In 1858 Mr. Robert Mallet published in the Reports of the British Association a list of several hundred works relating to earthquakes. Sixty-five of these works are to be found in the British Museum. So far as literature is concerned, earthquakes have received as much attention in the East as in the West. In China there are many works treating on earthquakes, and the attention which these phenomena received may be judged of from the fact that in A.D. 136 the Government appointed a commission to inquire into the subject. Even the isolated empire of Japan can boast of at least sixty-five works on earthquakes, seven of which are earthquake calendars, and twenty-three earthquake monographs.[2] Besides those treating especially of earthquakes, there are innumerable references to such disturbances in various histories, in the transactions of learned societies, and in periodicals. To attempt to give a complete catalogue of even the books which have been written would be to enter on a work of compilation which would occupy many years, and could never be satisfactorily finished.

In the 'Philosophical Transactions of the Royal Society,' which were issued in the eighteenth century, there are about one hundred and eighty separate communications on earthquakes; and in the 'Gentleman's Magazine' for 1755 there are no less than fifty notes and articles on the same subject. The great interest shown in earthquakes about the years 1750-60 in England, was chiefly due to the terrible calamity which overtook Lisbon in 1755, and to the fact that about this time several shocks were experienced in various parts of the British Islands. In 1750, which may

[1] *Mémoires de l'Académie Imp. de Dijon*, vols. xiv. and xv., 2nd Series, 1855-56.

[2] *Trans. Seis. Soc. of Japan*, vol. iii. p 65.

be described as the earthquake year of Britain, 'a shock
was felt in Surrey on March 14; on the 18th of the
same month the whole of the south-west of England was
disturbed. On April 2, Chester was shaken; on June 7,
Norwich was disturbed; on August 23, the inhabitants
of Lancashire were alarmed; and on September 30 ludicrous and alarming scenes occurred in consequence of
a shock having been felt during the hours of Divine
service, causing the congregations to hurry into the open
air.' [1]   As might be expected, these occurrences gave rise
to many articles and notes directing attention to the
subject of earthquakes.

Seismic literature has not, however, at all times been
a measure of seismic activity : thus, in Japan, the earthquake records for the twelfth and sixteenth centuries
scarcely mention any shocks. At first sight it might be
imagined that this was owing to an absence of earthquakes;
but it is sufficiently accounted for by the fact that at that
time the country was torn with civil war, and matters
more urgent than the recording of natural phenomena
engaged the attention of the inhabitants. Professor Rockwood, who has given so much attention to seismic disturbances in America, tells us that during the recent
contest between Chili and Peru a similar intermission is
observable. We see, therefore, that an absence of records
does not necessarily imply an absence of the phenomena
to be recorded.

Perhaps the earliest existing records of earthquakes
are those which occur in the Bible. The first of these,
which we are told occurred in Palestine, was in the reign
of Ahab (B.C. 918–897).[2]   One of the most terrible earthquakes mentioned in the Bible is that which took place
in the days of Uzziah, king of Judah (B.C. 811–759),

---

[1] *Gentleman's Magazine*, 1753.   [2] 1 Kings xix. 11, 12.

which shook the ground and rent the Temple. The awful character of this, and the deep impression produced on men's minds, may be learned from the fact that the time of its occurrence was subsequently used as an epoch from which to reckon dates.

The writings of Herodotus, Pliny, Livy, &c., &c., show the interest which earthquakes attracted in early ages. These writers chiefly devoted themselves to references and descriptions of disastrous shocks, and to theories respecting the cause of earthquakes.

The greater portion of the Japanese notices of earthquakes is simply a series of anecdotes of events which took place at the time of these disasters. We also find references to superstitious beliefs, curious occurrences, and the apparent connection between earthquake disturbances and other natural phenomena. In these respects the literature of the East closely resembles that of the West. The earthquake calendars of the East, however, form a class of books which can hardly be said to find their parallel in Europe; [1] while, on the other hand, the latter possesses types of books and pamphlets which do not appear to have a parallel elsewhere. These are the more or less theological works—'Moral Reflections on Earthquakes,' 'Sermons' which have been preached on earthquakes, 'Prayers' which have been appointed to be read. [2]

Speaking generally, it may be said that the writings of the ancients, and those of the Middle Ages, down to the commencement of the nineteenth century, tended to the propagation of superstition and to theories based on

[1] 'Notes on the Great Earthquake of Japan.' J. Milne, *Trans. Seis. Soc. of Japan*, vol. iii.

[2] See Mallet's List of Works on Earthquakes, *Report of the British Association*, 1858, p. 107.

speculations with few and imperfect facts for their foundation.

Among the efforts which have been made in modern times to raise seismology to a higher level, is that of Professor Perrey, of Dijon, who commenced in 1840 a series of extensive catalogues embracing the earthquakes of the world. These catalogues enabled Perrey, and subsequently Mallet in his reports to the British Association, to discuss the periodicity of earthquakes, with reference to the seasons and to other phenomena, in a more general manner than it had been possible for previous workers to accomplish. The facts thus accumulated also enabled Mallet to discuss earthquakes in general, and the various phenomena which they present were sifted and classified for inspection. Another great impetus which observational seismology received was Mr. Mallet's report upon the Neapolitan earthquake of 1857, in which new methods of seismic investigation were put forth. These have formed the working tools of many subsequent observers, and by them, as well as by his experiments on artificially produced disturbances, Mallet finally drew the study of earthquakes from the realms of speculation by showing that they, like other natural phenomena, were capable of being understood and investigated.

In addition to Perrey and Mallet, the nineteenth century has produced many writers who have taken a considerable share in the advancement of seismology. There are the catalogues of Von Hoff, the observations of Humboldt, the theoretical investigations of Hopkins, the monographs of Schmidt, Seebach, Lasaulx, and others; the books of Fuchs, Credner, Vogt, Volger; the records and observations of Palmieri, Bertelli, Rossi, and other Italian observers. To these, which are only a few out of a long list of names, may be added the publications of the

Commission appointed for the observation of earthquakes by the Natural History Society of Switzerland, and the volumes which have been published by the Seismological Society of Japan.

Before concluding this chapter it will be well to define a few of the more ordinary terms which are used in describing earthquake phenomena. It may be observed that the English word *earthquake*, the German *erdbeben*, the French *tremblement de terre*, the Spanish *terremoto*, the Japanese *jishin* &c., all mean, when literally translated, *earth-shaking*, and are popularly understood to mean a sudden and more or less violent disturbance.

Seismology (σεισμός an earthquake, λόγος a discourse) in its simplest sense means the study of earthquakes. To be consistent with a Greek basis for seismological terminology, some writers have thrown aside the familiar expression ' earthquake,' and substituted the awkward word ' seism.'

The source from which an earthquake originates is called the ' origin,' ' focal cavity,' or ' centrum.'

The point or area on the surface of the ground above the origin is called the ' epicentrum.' The line joining the centrum and epicentrum is called the ' seismic vertical.'

The radial lines along which an earthquake may be propagated from the centrum are called ' wave-paths.'

The angle which a wave-path, where it reaches the surface of the earth, makes with that surface is called the ' angle of emergence ' of the wave. This angle is usually denoted by the letter e.

As the result of a simple explosion at a point in a homogeneous medium, we ought, theoretically, to obtain at points on the surface of the medium equidistant from the epicentrum, equal mechanical effects. These points will lie on circles called 'isoseismic' or 'coseismic' circles. The area included between two such circles is an 'isoseismic area.' In nature, however, isoseismic lines are seldom circles. Elliptical or irregular curves are the common forms.

The isoseismic area in which the greatest disturbance has taken place is called the 'meizoseismic area.' Seebach calls the lines enclosing this area 'pleistoseists.'

These last-mentioned lines are wholly due to Mallet and Seebach.

Many words are used to distinguish different kinds of earthquakes from each other. All of these appear to be very indefinite and to depend upon the observer's feelings, which, in turn, depend upon his nervous temperament and his situation.

In South America small earthquakes, consisting of a series of rapidly recurring vibratory movements not sufficiently powerful to create damage, are spoken of as *trembelores.*

The *terremotos* of South America are earthquakes of a destructive nature, in which distinct shocks are perceptible. It may be observed that shocks which at one place would be described as *terremoto* would at another and more distant place probably be described as *trembelores.*

The *succussatore* are the shocks where there is considerable vertical motion. The terrible shock of Reobamba (February 4, 1797), which is said to have thrown corpses from their graves to a height of 100 feet, was an earthquake of this order.

The *vorticossi* are shocks which have a twisting or rotatory motion.

Another method of describing earthquakes would be

to refer to instrumental records. When the vibrations of the ground have only been along the line joining the observer and the epicentrum, the disturbance might be called 'euthutropic.' A disturbance in which the prominent movements are *transverse* to the above direction might be called 'diagonic.' If motions in both of these directions occur in the records, the shock might be said to be 'diastrophic.' If there be much vertical movement, the shock might be said to be 'anaseismic.' Some disturbances could only be described by using two or three of these terms.

CHAPTER II.

SEISMOMETRY.

Nature of earthquake vibrations—Many instruments called seismometers only seismoscopes—Eastern seismoscopes, columns, projection seismometers—Vessels filled with liquid—Palmieri's mercury
tubes—The ship seismoscope—The cacciatore—Pendulum instruments of Kreil, Wagner, Ewing, and Gray—Bracket seismographs—
West's parallel motion instrument—Gray's conical pendulums,
rolling spheres, and cylinders—Verbeck's ball and plate seismograph
—The principle of Perry and Ayrton—Vertical motion instruments
—Record receivers — Time-recording apparatus — The Gray and
Milne seismograph.

BEFORE we discuss the nature of earthquake motion, the
determination of which has been the aim of modern seismological investigation, the reader will naturally look for
an account of the various instruments which have been
employed for recording such disturbances.   A description
of the earthquake machines which have been used even in
Japan would form a bulky volume.   All that we can do,
therefore, is to describe briefly the more prominent features of a few of the more important of these instruments.
In order that the relative merits of these may be better
understood, we may state generally that modern research
has shown a typical earthquake to consist of a series of
small tremors succeeded by a shock, or series of shocks,
separated by more or less irregular vibrations of the
ground.   The vibrations are often both irregular in period

and in amplitude, and they have a duration of from a few seconds to several minutes. We will illustrate the records of actual earthquakes in a future chapter, but in the meantime the idea that an earthquake consists of a single shock must be dismissed from the imagination.

To construct an instrument which at the time of an earthquake shall move and leave a record of its motion, there is but little difficulty. Contrivances of this order are called *seismoscopes*. If, however, we wish to know the period, extent, and direction of each of the vibrations which constitutes an earthquake, we have considerable difficulty. Instruments which will in this way measure or write down the earth's motions are called *seismometers* or *seismographs*.

Many of the elaborate instruments supplemented with electro-magnetic and clockwork arrangements are, when we examine them, nothing more than elaborate seismoscopes which have been erroneously termed seismographs.

The only approximations to true seismographs which have yet been invented are without doubt those which during the past few years have been used in Japan. It would be a somewhat arbitrary proceeding, however, to classify the different instruments as seismoscopes, seismometers, and seismographs, as the character of the record given by certain instruments is sometimes only seismoscopic, whilst at other times it is seismometric, depending on the nature of the disturbance. Many instruments, for instance, would record with considerable accuracy a single sudden movement, but would give no reliable information regarding a continued shaking.

Eastern Seismoscopes.— The earliest seismoscope of which we find any historical record is one which owes its origin to a Chinese called Chôko. It was invented in the year A.D. 136. A description is given in the Chinese

history called 'Gokanjo,' and the translation of this description runs as follows :—

    ' In the first year of Yōka, A.D. 136, a Chinese called
Chôko invented the seismometer shown in the accompanying drawing. This instrument consists of a spherically
formed copper vessel, the diameter of which is eight feet.
It is covered at its top, and in form resembles a wine-

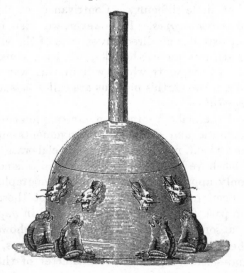

FIG. 1.

bottle. Its outer part is ornamented by the figures of
different kinds of birds and animals, and old peculiar-
looking letters. In the inner part of this instrument a
column is so suspended that it can move in eight directions. Also, in the inside of the bottle, there is an
arrangement by which some record of an earthquake is
made according to the movement of the pillar. On the
outside of the bottle there are eight dragon heads, each

of which holds a ball in its mouth. Underneath these heads there are eight frogs so placed that they appear to watch the dragon's face, so that they are ready to receive the ball if it should be dropped. All the arrangements which cause the pillar to knock the ball out of the dragon's mouth are well hidden in the bottle.'

' When an earthquake occurs, and the bottle is shaken, the dragon instantly drops the ball, and the frog which receives it vibrates vigorously ; any one watching this instrument can easily observe earthquakes.'

With this arrangement, although one dragon may drop a ball, it is not necessary for the other seven dragons to drop their balls unless the movement has been in all directions ; thus we can easily tell the direction of an earthquake.

' Once upon a time a dragon dropped its ball without any earthquake being observed, and the people therefore thought the instrument of no use, but after two or three days a notice came saying that an earthquake had taken place at Rōsei. Hearing of this, those who doubted the use of this instrument began to believe in it again. After this ingenious instrument had been invented by Chōko, the Chinese Government wisely appointed a secretary to make observations on earthquakes.'

Not only is this instrument of interest on account of its antiquity, but it is also of interest on account of the close resemblance it bears to many of the instruments of modern times.

Another earthquake instrument also of Eastern origin is the magnetic seismoscope of Japan.

On the night of the destructive earthquake of 1855, which devastated a great portion of Tokio, the owner of a spectacle shop in Asakusa observed that a magnet dropped

some old iron nails and keys which had been attached to
it. From this occurrence the owner thought that the
magnet had, in consequence of its age, lost its powers.
About two hours afterwards, however, the great earthquake took place, after which the magnet was observed to
have regained its powers. This occurrence led to the
construction of the seismoscope, which is illustrated in a
book called the ' Ansei-Kembun-Roku,' or a description
of the earthquake of 1855, and examples of the instrument are still to be seen in Tokio. These instruments
consist of a piece of magnetic iron ore, which holds up a
piece of iron like a nail. This nail is connected, by means
of a string, with a train of clockwork communicating with
an alarum. If the nail falls a catch is released and the
clockwork set in motion, and warning given by the ring-
ing of a bell. It does not appear that this instrument
has ever acted with success.

Columns.—One of the commonest forms of seismoscope, and one which has been very widely used, consists
of a round column of wood, metal, or other suitable
material, placed, with its axis vertical, on a level plane,
and surrounded by some soft material such as loose sand
to prevent it rolling should it be overturned. The fall
of such a column indicates that a shaking or shock has
taken place. Attempts have been made by using a
number of columns of different sizes to make these indications seismometric, but they seldom give reliable
information either as to intensity or direction of shock.
The indications as to intensity are vitiated by the fact
that a long-continued gentle shaking may overturn a
column which would stand a very considerable sudden
shock, while the directions in which a number of columns
fall seldom agree owing to the rotational motion imparted
to them by the shaking. Besides, the direction of motion

of the earthquake seldom remains in the same azimuth throughout the whole disturbance.

An extremely delicate, and at the same time simple form of seismoscope may be made by propping up strips of glass, pins, or other easily overturned bodies against suitably placed supports. In this way bodies may be arranged, which, although they can only fall in one direction, nevertheless fall with far less motion than is necessary to overturn any column which will stand without lateral support.

Projection Seismometers.—Closely related to the seismoscopes and seismometers which depend on the overturning of bodies. Mallet has described two sets of apparatus whose indications depend on the distance to which a body is projected. In one of these, which consisted of two similar parts arranged at right angles, two metal balls rest one on each side of a stop at the lower part of two inclined $\diagdown\diagup$ like troughs. In this position each of the balls completes an electric circuit. By a shock the balls are projected or rolled up the troughs, and the height to which they rise is recorded by a corresponding interval in the break of the circuits. The vertical component of the motion is measured by the compression of a spring which carries the table on which this arrangement rests. In the second apparatus two balls are successively projected, one by the forward swing, and the other by the backward swing of the shock. Attached to them are loose wires forming terminals of the circuits. They are caught in a bed of wet sand in a metal trough forming the other end of the circuit. The throw of the balls as measured in the sand, and the difference of time between their successive projections as indicated by special contrivances connected with the closing of the circuits, enables the observer to calculate the direction of the wave of shock,

C

its velocity, and other elements connected with the disturbance. It will be observed that the design of this apparatus assumes the earthquake to consist of a distinct isolated shock.

Oldham, at the end of his account of the Cachar earthquake of 1869, recommends the use of an instrument based on similar principles. In his instrument four balls like bullets are placed in notches cut in the corners of the upper end of a square stake driven into the ground.

Vessels filled with liquid.—Another form of simple seismoscope is made by partially filling a vessel with liquid. The height to which the liquid is washed up the side of the vessel is taken as an indication of the intensity of the shock, and the line joining the points on which maximum motion is indicated, is taken as the direction of the shock. If earthquakes all lasted for the same length of time, and consisted of vibrations of the same period, such instruments might be of service. These instruments have, however, been in use from an early date. In 1742 we find that bowls of water were used to measure the earthquakes which in that year alarmed the inhabitants of Leghorn. About the same time the Rev. S. Chandler, writing about the shock at Lisbon, tells us that earthquakes may be measured by means of a spherical bowl about three or four feet in diameter, the inside of which, after being dusted over with Barber's puff, is filled very gently with water. Mallet, Babbage, and De la Bêche have recommended the same sort of contrivance, but, notwithstanding, it has justly been criticised as 'ridiculous and utterly impracticable.' [1]

An important portion of Palmieri's well-known instrument consists of horizontal tubes turned up at the ends

[1] *Quarterly Review*, vol. lxiii. p. 61.

and partially filled with mercury. To magnify the motion of the mercury, small floats of iron rest on its surface. These are attached by means of threads to a pulley provided with indices which move in front of a scale of degrees. We thus read off the intensity of an earthquake as so many degrees, which means so many millimetres of washing up and down of mercury in a tube. The direction of movement is determined by the azimuth of the tube which gives the maximum indication, several tubes being placed in different azimuths.

This form of instrument appears to have been suggested by Mallet, who gives an account of the same in 1846. Inasmuch as the rise and fall of the mercury in such tubes depend on its depth and on the period of the earthquake together with its duration, we see that although the results obtained from a given instrument may give us means of making approximate comparisons as to the relative intensity of various earthquakes, it is very far from yielding any absolute measurement.

Another method which has been employed to magnify and register the motions of liquid in a vessel has been to float upon its surface a raft or ship from which a tall mast projected. By a slight motion of the raft, the top of the mast vibrated through a considerable range. This motion of the mast as to direction and extent was then recorded by suitable contrivances attached to the top of the mast.

A very simple form of liquid seismoscope consists of a circular trough of wood with notches cut round its side. This is filled with mercury to the level of the notches. At the time of an earthquake the maximum quantity of mercury runs over the notches in the direction of greatest motion. This instrument, which has long been used in Italy, is known as a Cacciatore, being named after its inventor. It is a prominent feature in the collection of apparatus forming the well-known seismograph of Palmieri.

Pendulum instruments.—Mallet speaks of pendulum seismoscopes and seismographs as 'the oldest probably of seismometers long set up in Italy and southern Europe.' In 1841 we find these being used to record the earthquake disturbances at Comrie in Scotland.

These instruments may be divided into two classes: first, those which at the time of the shock are intended to swing, and thus record the direction of movement; and second, those which are supposed to remain at rest and thus provide 'steady points.'

To obtain an absolutely 'steady point' at the time of an earthquake, has been one of the chief aims of all recent seismological investigations.

With a style or pointer projecting down from the steady point to a surface which is being moved backward and forward by the earth, such a surface has written upon it by its own motions a record of the ground to which it is attached. Conversely, a point projecting upwards from the moving earth might be caused to write a record on the body providing the steady point, which in the class of instruments now to be referred to is supposed to be the bob of a pendulum. It is not difficult to get a pendulum which will swing at the time of a moderately strong earthquake, but it is somewhat difficult to obtain one which will not swing at such a time. During the past few years, pendulums varying between forty feet in length and carrying bobs of eighty pounds in weight, and one-eighth of an inch in length, and carrying a gun-shot, have been experimented with under a great variety of circumstances. Sometimes the supports of these pendulums have been as rigid as it is possible to make a structure from brick and mortar, and at other times they have intentionally been made loose and flexible. The indices which wrote the motions of these pendulums have been as various as the pendulums themselves. A small needle sliding vertically through two small holes, and resting its lower end on a surface of smoked glass, has on account of its small amount of friction been perhaps one of the favourite forms of recording pointers.

The free pendulums which have been employed, and which were intended to swing, have been used for two purposes : first, to determine the direction of motion from the direction of swing, and second, to see if an approximation to the period of the earth's motion could be obtained by discovering the pendulum amongst a series of different lengths which was set in most violent motion, this probably being the one which had its natural period of swing the most nearly approximating to the period of the earthquake oscillations.

Inasmuch as all pendulums when swinging have a tendency to change the plane of their oscillation, and also as we now know that the direction of motion during an earthquake is not always constant, the results usually obtained with these instruments respecting the direction of the earth's motion have been unsatisfactory. The results which were obtained by series of pendulums of different lengths were, for various reasons, also unsatisfactory.

Of pendulums intended to provide a steady point, from which the relative motion of a point on the earth's surface could be recorded, there has been a great variety. One of the oldest forms consisted of a pendulum with a style projecting downwards from the bob so as to touch a bed of sand. Sometimes a concave surface was placed beneath the pendulum, on which the record was traced by means of a pencil. Probably the best form was that in which a needle, capable of sliding freely up and down, marked the relative horizontal motion of the earth and the pendulum bob on a smoked glass plate.

It generally happens that at the time of a moderately severe earthquake the whole of these forms of apparatus are set in motion, due partly to the motion of the point of support of the pendulum, and partly to the friction of the writing point on the plate.

Among these pendulums may be mentioned those of Cavallieri, Faura, Palmieri, Rossi, and numerous others. It is possible that the originators of some of these pendulums may have intended that they should record by swinging. If this is so, then so far as the determination of the actual nature of earthquake motion is concerned, they belong to a lower grade of apparatus than that in which they are here included.

A great improvement in pendulum apparatus is due to Mr. Thomas Gray of Glasgow, who suggested applying so much frictional resistance to the free swing of a pendulum that for small displacements it became 'dead beat.' By carrying out this suggestion, pendulum instruments were raised to the position of seismographs. The manner of applying the friction will be understood from the following description of a pendulum instrument which is also provided with an index which gives a magnification of the motion of the earth.

B B B B is a box 113 cm. high and 30 cm. by 18 cm. square. Inside this box a lead ring R, 17 cm. in diameter and 3 cm. thick, is suspended as a pendulum from the screw s. This screw passes through a small brass plate P P, which can be moved horizontally over a hole in the top of the box. These motions in the point of suspension allow the pendulum to be adjusted.

Projecting over the top of the pendulum there is a

wooden arm W carrying two sliding pointers H H, resting on a glass plate placed on the top of the pendulum. These pointers are for the purpose of giving the frictional resistance before referred to. If this friction plate is smoked, the friction pointers will write upon it records of *large* earthquakes independently of the records given by the proper index, which only gives satisfactory records in the case of shocks of ordinary intensity. Crossing the inside of the pendulum R there is a brass bar perforated with a small conical hole at M. A stiff wire passes through M and forms the upper portion of the index I, the lower portion of which is a thin piece of bamboo. Fixed

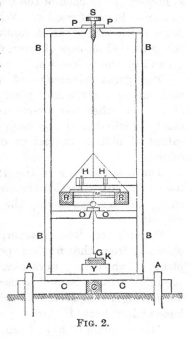

FIG. 2.

upon the wire there is a small brass ball which rests on the upper side of a second brass plate also perforated with a conical hole, which plate is fixed on the bar O O crossing the box.

    If at the time of an earthquake the upper part of the index I remains steady at M, then by the motion at O, the lower end of the index which carries a sliding needle at G, will magnify the motion of the earth in the ratios M O : O G.   In this instrument O G is about 17 cm.

    The needle G works upon a piece of smoked glass.   In

order to bring the glass into contact with the needle without disturbance, the glass is carried on a strip of wood K, hinged at the back of the box, and propped up in front by a loose block of wood Y. When Y is removed the glass drops down with K out of contact with the needle. The box is carried on bars of wood C C, which are fixed to the ground by the stakes A A.

The great advantage of a pendulum seismograph working on a stationary plate is, that the record shows at once whether the direction of motion has been constant, or whether it has been variable. The maximum extent of motion in various directions is also easily obtained.

The disadvantage of the instrument is, that at the time of a large earthquake, owing perhaps to a slight swing in the pendulum, the records may be unduly magnified.

On such occasions, however, fairly good records may be obtained from the friction pointers, provided that the plates on which they work have been previously smoked. It might perhaps be well to use two of these instruments, one having a comparatively high frictional resistance, and hence 'dead beat' for large displacements.

Many attempts have been made to use a pendulum seismograph in conjunction with a record-receiving surface, which at the time of the earthquake should be kept in motion by clockwork. In this way it was hoped to separate the various vibrations of the earthquake, and thus avoid the greater or less confusion which occurs when the index of the pendulum writes its backward and forward motion on a stationary plate. Hitherto all attempts in this direction, in which a single multiplying index was used, have been unsuccessful because of the moving plate dragging the index in the direction of its motion for a short distance, and then allowing it to fall back towards its normal position.

In connection with this subject we may mention the pendulum seismographs of Kreil, Wagener, Ewing, and Gray.

In the bob of Kreil's pendulum there was clockwork, which caused a disc on the axis of the pendulum to continuously rotate. On this continually revolving surface a style fixed to the earth traced an unbroken circle. At the time of an earthquake, by the motion of the style, the circle was to be broken and lines drawn. The number and length of these lines were to indicate the length and intensity of the disturbance.

Gray's pendulum consisted of a flat heavy disc carrying on its upper surface a smoked glass plate. This, which formed the bob of the pendulum, was supported by a pianoforte steel wire. When set ready to receive an earthquake, the wire was twisted and the bob held by a catch so arranged that at the time of the earthquake the catch was released, and the bob of the pendulum allowed to turn slowly by the untwisting of the supporting wire. Resting on the surface of this rotating disc were two multiplying indices arranged to write the earth's motions as two components.

In the instruments of Wagener and Ewing, the clockwork and moving surface do not form part of the pendulum, but rest independently on a support rigidly attached to the earth. In Wagener's instrument one index only is used, while in Ewing's two are used for writing the record of the motion.

A difficulty which is apparent in all pendulum machines is that when the bob of such a pendulum is deflected it tends to fall back to its normal position. To make a pendulum perfect it therefore requires some compensating arrangement, so that the pendulum, for small displacements, shall be in neutral equilibrium, and the errors due to swinging shall be avoided.

Several methods have been suggested for making the bob of an ordinary pendulum astatic for small displacements. One method proposed by Gray consists in fixing in the bob of a pendulum a circular trough of liquid, the curvature of this trough having a proper form. Another method which was suggested, was to attach a vertical spiral spring to a point in the axis of the pendulum a little below the point of suspension, and to a fixed point above it, so that when the pendulum is deflected it would introduce a couple.

Professor Ewing has suggested an arrangement so that the bob of the pendulum shall be partly suspended by a stretched spiral spring, and at the same time shall be partly held up from below by a vertically placed strut, the weight carried by the strut being to the weight carried by the spring in the ratio of their respective lengths. As to how these arrangements will act when carried into practice yet remains to be seen.

Another important class of instruments are *inverted pendulums*. These are vertical springs made of metal or wood loaded at their upper end with a heavy mass of metal. An arrangement of this sort, provided at its upper end with a pencil to write on a concave surface, was employed in 1841 to register the earthquakes at Comrie in Scotland. In Japan they were largely employed in series, each member of a series having a different period of vibration. The object of these arrangements was to determine which of the pendulums, with a given earthquake, recorded the greatest motion, it being assumed that the one which was thrown into the most violent oscillation would be the one most nearly approximating with the period of the earthquake. The result of these experiments showed that it was usually those with a slow period of vibration which were the most disturbed.

Bracket Seismographs.—A group of instruments of recent origin which have done good work, are the bracket seismographs. These instruments appear to have been independently invented by several investigators: the germ from which they originated probably being the well-known horizontal pendulum of Professor Zöllner. In Japan they were first employed by Professor W. S. Chaplin. Subsequently they were used by Professor Ewing and Mr. Gray. They consist essentially of a heavy weight supported at the extremity of a horizontal bracket which is free to turn on a vertical axis at its other end. When the frame carrying this axis is moved in any direction excepting parallel to the length of the gate-like bracket, the weight causes the bracket to turn round a line known as the instantaneous axis of the bracket corresponding to this motion of the fixed axis. Any point in this line may therefore be taken as a steady point for motions at right angles to the length of the supporting bracket. Two of these instruments placed at right angles to each other have to be employed in conjunction, and the motion of the ground is written down as two rectangular components. In Professor Ewing's form of the instrument, light prolongations of the brackets form indices which give magnified representations of the motion, and the weights are pivoted round a vertical axis through their centre.

In the accompanying sketch B is a heavy weight pivoted at the end of a small bracket C A K, which bracket is free to turn on a knife-edge, K, above, and a pivot A, below, in the stand S. At the time of an earthquake B remains steady, and the index P, forming a continuation

of the bracket, magnifies the motion of the stand, in the
ratio of A C : C N.

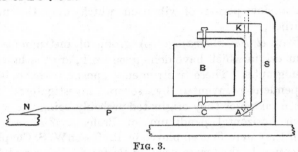

In an instrument called a double-bracket seismograph,
invented by Mr. Gray, we have two brackets hinged to
each other, and one of them to a fixed frame. The planes
of the two brackets are placed at right angles, so as to
give to a heavy mass supported at the end of the outer
bracket two degrees of horizontal freedom.

In all bracket machines, especially those which carry
a pivoted weight, it is doubtful whether the weight provides a truly steady point relatively to the plate on which
the record is written for motion parallel to the direction
of the arm.

Parallel-motion Instrument.—A machine which writes
its record as two components, and which promises great
stability, is one suggested by
Professor C. D. West. Like
the bracket machines it consists of two similar parts
placed at right angles to each
other, and is as follows: A
bar of iron A is suspended
from both sides on pivots at

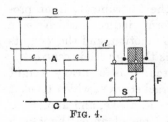

FIG. 4.

c c by a system of light arms hinging with each other at

the black dots, between the upper and lower parts of the rigid frame B C. The arms are of such a length that for small displacements parallel to the length of the bar, c c practically move in a straight line, and the bar is in neutral equilibrium. A light prolongation of the bar d works the upper end of the light index e, passing as a universal joint through the rigid support F. A second index e' from the bar at right angles also passes through F. The multiplying ends of these indices are coupled together to write a resultant motion on a smoked glass plate s.

Conical Pendulums.—Another group of instruments which have also yielded valuable records are the conical pendulum seismographs. The idea of using the bob of a conical pendulum to give a steady point in an earthquake machine was first suggested and carried into practice by Mr. Gray. The seismograph as employed consists of a pair of conical pendulums hung in planes at right angles to each other. The bob of each of these pendulums is fixed a short distance from the end of a light lever, which forms the writing index, the short end resting as a strut against the side of a post fixed in the earth. The weight is carried by a thin wire or thread, the upper end of which is attached to a point vertically above the fixed end of the lever.

Rolling Spheres and Cylinders.—After the conical pendulum seismographs, which claim several important advantages over the bracket machines, we come to a group of instruments known as rolling sphere seismographs. Here, again, we have a class of instruments for the various forms of which we are indebted to the ingenuity of Mr. Gray.

The general arrangement and principle of one of these instruments will be readily understood from the

accompanying figure. s is a segment of a large sphere
with a centre near c. Slightly below this centre a heavy
weight B, which may be a lead ring, is pivoted. At the
time of an earthquake c is steady, and the earth's motions
are magnified by the pointer C A N in the proportion of
C A : A N. The working of this pointer or index is similar
to that of the pointer in the pendulum.

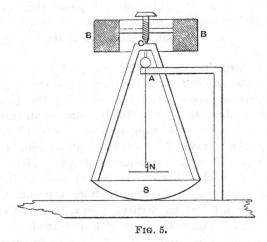

FIG. 5.

Closely connected with the rolling sphere seismographs, are Gray's rolling cylinder seismographs.

These are two cylinders resting on a surface plate
with their axes at right angles to each other. Near to
the highest point in each of these cylinders, this point
remaining nearly steady when the surface plate is moved
backwards and forwards, there is attached the end of a
light index. These indices are again pivoted a short
distance from their ends on axes connected with the
surface plate. In order that the two indices may be
brought parallel, one is cranked at the second pivot.

Ball and Plate Seismograph. — Another form of seismograph, which is closely related to the two forms of apparatus just described, is Verbeck's ball and plate seismograph. This consists of a surface plate resting on three hard spheres, which in turn rests upon a second surface plate. When the lower plate is moved, the upper one tends to remain at rest, and thus may be used as a steady mass to move an index.

The Principle of Perry and Ayrton.—An instrument which is of interest from the scientific principle it involves is a seismograph suggested by Professors Perry and Ayrton, who propose to support a heavy ball on three springs, which shall be sufficiently stiff to have an exceedingly quick period of vibration. By means of pencils attached to the ball by levers, the motions of the ball are to be recorded on a moving band of paper. The result would be a record compounded of the small vibrations of the springs superimposed on the larger, slower, wave-like motions of the earthquake, and, knowing the former of these, the latter might be separated by analysis. Although our present knowledge of earthquake motion indicates that the analysis of such a record would often present us with insuperable difficulties, this instrument is worthy of notice on account of the novelty of the principle it involves, which, the authors truly remark, has in seismometry been a ' neglected ' one.

Instruments to record Vertical Motion.—The instruments which have been devised to record vertical motion are almost as numerous as those which have been devised to record horizontal motion. The earliest form of instrument employed for this purpose was a spiral spring stretched by weight, which, on account of its inertia, was supposed at the time of a shock to remain steady. No satisfactory results have ever been obtained from such instruments, chiefly on account of the inconvenience in making a spring sufficiently long to allow of enough elongation to give a long period of vibration. Similar remarks may be applied to the horizontally placed elastic rods, one end of which is fixed to a wall, whilst the opposite end is loaded with a weight. Such contrivances, furnished with pencil on the weight to write a record upon a vertical surface, were used in 1842 at Comrie, and we see the same principle applied in a portion of Palmieri's apparatus. Contrivances like these neither give us the true amplitude of the vertical motion, insomuch as they are readily set in a state of oscillation; nor do they indicate the duration of a disturbance, for, being once set in motion, they continue that motion in virtue of their inertia long after the actual earthquake has ceased. They can only be regarded as seismoscopes.

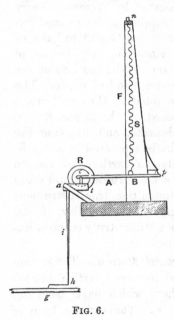

FIG. 6.

The most satisfactory instrument which has yet been devised for recording vertical motion is Gray's horizontal-lever spring seismograph.

This instrument will be better understood from the accompanying sketch. A vertical spring s is fixed at its upper end by means of a nut n, which rests on the top of the frame F, and serves to raise or lower the spring through a short distance as a last adjustment for the position of the cross-arm A. The arm A rests at one end on two sharp points, p, one resting in a conical hole and the other in a v-slot; it is supported at B by the spring s, and is weighted at C with a lead ring R. Over a pin at the point C a stirrup of thread is placed which supports a small trough, t. The trough t is pivoted at a, has attached to it the index i (which is hinged by means of a strip of tough paper at h, and rests through a fine pin on the glass plate g), and is partly filled with mercury.

Another method of obtaining a steady point for vertical motion is that of Dr. Wagener, who employs a buoy partly immersed in a vessel of water. This was considerably improved upon by Mr. Gray, who suggested the use of a buoy, which, with the exception of a long thin style, was completely sunk.

Among the other forms of apparatus used to record vertical motion may be mentioned vessels provided with india-rubber or other flexible bottoms, and partially filled with water or some other liquid. As the vessel is moved up and down, the bottom tends to remain behind and provides a more or less steady point. Pivoted to this is a light index, which is again pivoted to a rigid frame in connection with the earth. Instruments of this description have yielded good records.

Record Receivers.—A large number of earthquake machines having been referred to, it now remains to consider the apparatus on which they write their motions. The earlier forms of seismographs, as has already been indicated, recorded their movements in a bed of sand; others wrote their records by means of pencils on sheets of paper. Where we have seismographs which magnify the motion of the earth, it will be observed that methods like the above would involve great frictional resistances,

D

tending to cause motion in the assumed steady points of
the seismographs. One of the most perfect instruments
would be obtained by registering photographically the
motions of the recording index by the reflection of a ray
of light. Such an instrument would, however, be difficult
to construct and difficult to manipulate. One of the best
practical forms of registering apparatus is one in which
the record is written on a surface of smoked glass. This
can afterwards be covered with a coat of photographer's
varnish, and subsequently photographed by the 'blue process' so well known to engineers.

To obtain a record of all the vibrations of an earthquake it is necessary that the surface on which the seismograph writes should at the time of an earthquake be in
motion. Of record-receiving machines there are three
types. First, there are those which move continuously.
The common form of these is a circular glass plate like
an old form of chronograph, driven continuously by clockwork. On this the pointers of the seismograph rest and
trace over and over again the same circles. At the time
of an earthquake they move back and forth across the
circles, which are theoretically fine lines, and leave a
record of the earthquake. Instead of a circular plate, a
drum covered with smoked paper may be used, which,
after the earthquake, possesses the advantage, after un-
rolling, of presenting the record in a straight line, instead
of a record written round the periphery of a circle, as is
the case with the circular glass plates. Such records are
easily preserved, but they are more difficult to photograph.

The second form of apparatus is one which is set in
motion at the time of a shock. This may be a contrivance like one of those just described, or a straight
smoked glass plate on a carriage. By means of an elec-

trical or a mechanical contrivance called a 'starter,' of which many forms have been contrived, the earthquake is caused to release a detent and thus set in motion the mechanism which moves the record-receiver.

The great advantage of continuously-moving machines is that the beginning and end of the shock can usually be got with certainty, while all the uncertainty as to the action of the 'starter' is avoided. Self-starting machines have, of course, the advantage of simplicity and cheapness, while there is no danger of the record getting obliterated by the subsequent motion of the plate under the index.

Time-recording Apparatus.—Of equal importance with the instruments which record the motion of the ground, are those instruments which record the time at which such motion took place. The great value of time-records, when determining the origin from which an earthquake originates, will be shown farther on. The most important result which is required in connection with time observations, is to determine the interval of time taken by a disturbance in travelling from one point to another. On account of the great velocity with which these disturbances sometimes travel, it is necessary that these observations should be made with considerable accuracy. The old methods of adapting an apparatus to a clock which, when shaken, shall cause the clock to stop, are of little value unless the stations at which the observations are made are at considerable distances apart. This will be appreciated when we remember that the disturbance may possibly travel at the rate of a mile per second, that its duration at any station may often extend over a minute, and that one set of apparatus at one station may stop, perhaps, at the commencement of the disturbance, and the other near the end. A satisfactory time-taking apparatus will therefore require, not only the means for stopping a clock, but also a contrivance which, at the same instant that the clock is stopped, shall make a mark on a record which is being drawn by a seismograph. In this way we find out at which portion of the shock the time was taken.

Palmieri stops a clock in his seismograph by closing an electric circuit. Mallet proposes to stop a clock by

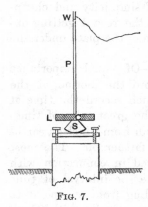

the falling of a column which is attached by a string to the pendulum of the clock. So long as the column is standing the string is loose and the pendulum is free to move; but when the column falls, the string is tightened and the pendulum is arrested. The difficulty which arises is to obtain a column that will fall with a slight disturbance. The best form of contrivance for causing a column to fall, and one which may also be used in drawing out a catch to relieve the machinery of a record-receiver, is shown in the accompanying sketch.

Fig. 7.

s is the segment of a sphere about 4·5 cm. radius, with a centre slightly above C. L is a disc of lead about 7 cm. in diameter resting upon the segment. Above this there is a light pointer, P, about 30 cm. long. On the top of the pointer a small cylinder of iron, W, is balanced, and connected by a string with the catch to be relieved. When the table on which W P S rests is shaken, rotation takes place near to C, the motion of the base S is magnified at the upper end of the pointer, and the weight is overturned. This catch may be used to relieve a toothed bar axled at one end, and held up above a pin projecting from the face of the pendulum bob. When this falls it catches the projecting pin and holds the pendulum.

Another way of relieving the toothed bar is to hold up the opposite end to that at which it is axled by resting it on the extremity of a horizontal wire fixed to the bob of a conical pendulum—for example, one of the indices of a conical pendulum seismograph. The whole of this apparatus, which may be constructed at the cost of a few pence, can be made small enough to go inside an ordinary clock case.

The difficulty which arises with all these clock-stopping arrangements is that it is difficult for observers situated at distant stations to re-start their clocks so that their difference in time shall be accurately known. Even if each observer is provided with a well-regulated chronometer, with which he can make comparisons, the rating of these instruments is for all ordinary persons an extremely troublesome operation.

In order to avoid this difficulty the author has of late years used a method of obtaining the time without stopping the clock. To do this a clock with a central seconds hand is taken, and the hour and minute hands are prolonged and bent out slightly at their extremities at right angles to the face, the hour hand being slightly the longest. Each hand is then tipped with a piece of soft material like cork, which is smeared with a glycerine ink. A light flat ring, with divisions in it corresponding to those on the face of the clock, is so arranged that at the time of a shock it can be quickly advanced to touch the inked pads on the hands of the clock and then withdrawn. This is accomplished by suitable machinery, which is relieved either by an electro-magnet or some other contrivance which will withdraw a catch. In this way an impression in the form of three dots is received on the disc, and the time known without either stopping or sensibly retarding the clock.

For ordinary observers, if a time-taker is not used in conjunction with a record-receiver, as good results as those obtained by ordinary clock-stopping apparatus are obtainable by glancing at an ordinary watch. Subsequently the watch by which the observation was made should be compared with some good time-keeper, and the local time at which the shock took place is then approximately known.

From what has now been said it will be seen that for a complete seismograph we require three distinct sets of apparatus—an apparatus to record horizontal motion, an apparatus to record vertical motion, and an apparatus to record time. The horizontal and vertical motions must be written on the same receiver, and if possible side by side, whilst the instant at which the time record is made a mark must be made on the edge of the diagram which is being drawn by the seismograph. Such a seismograph has been constructed and is now erected in Japan. It is illustrated in the accompanying diagram.

The Gray and Milne Seismograph.—In this apparatus two mutually rectangular components of the horizontal motion of the earth are recorded on a sheet of smoked paper wound round a drum, D, kept continuously in motion by clockwork, W, by means of two conical pendulum-seismographs, C. The vertical motion is recorded on the same sheet of paper by means of a compensated-spring seismograph, S L M B.

The time of occurrence of an earthquake is determined by causing the circuit of two electro-magnets to be closed by the shaking. One of these magnets relieves a mechanism, forming part of a time-keeper, which causes the dial of the timepiece to come suddenly forwards on the hands and then move back to its original position.

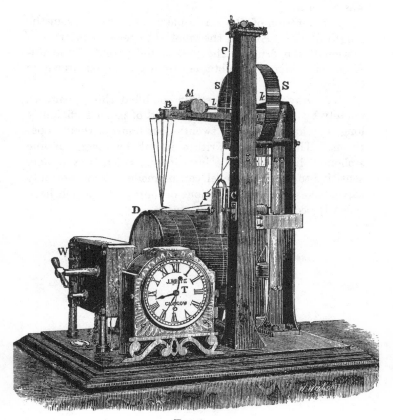

FIG. 8.

The hands are provided with ink-pads, which mark their positions on the dial, thus indicating the hour, minute, and second when the circuit was closed. The second electro-magnet causes a pointer to make a mark on the paper receiving the record of the motion. This mark indicates the part of the earthquake at which the circuit was closed.

The duration of the earthquake is estimated from the length of the record on the smoked paper and the rate of motion of the drum. The nature and period of the different movements are obtained from the curves drawn on the paper.

Mr. Gray has since greatly modified this apparatus, notably by the introduction of a band of paper sufficiently long to take a record for twenty-four hours without repetition. The record is written in ink by means of fine siphons. In this way the instrument, which is extremely sensitive to change of level, can be made to show not only earthquakes, but the pulsations of long period which have recently occupied so much attention.

CHAPTER III.

EARTHQUAKE MOTION DISCUSSED THEORETICALLY.

Ideas of the ancients (the views of Travagini, Hooke, Woodward, Stuckeley, Mitchell, Young, Mallet)—Nature of elastic waves and vibrations—Possible causes of disturbance in the Earth's crust—The time of vibration of an earth particle—Velocity and acceleration of a particle—Propagation of a disturbance as determined by experiments upon the elastic moduli of rocks—The intensity of an earthquake—Area of greatest overturning moment—Earthquake waves—Reflexion, refraction, and interference of waves—Radiation of a disturbance.

Ideas of Early Writers.—One of the first accounts of the varieties of motion which may be experienced at the time of an earthquake is to be found in the classification of earthquakes given by Aristotle.[1] It is as follows:—

1. Epiclintæ, or earthquakes which move the ground obliquely.

2. Brastæ, with an upward vertical motion like boiling water.

3. Chasmatiæ, which cause the ground to sink and form hollows.

4. Rhectæ, which raise the ground and make fissures.

5. Ostæ, which overthrow with one thrust.

6. Palmatiæ, which shake from side to side with a sort of tremor.

From the sixth group in this classification we see that

[1] *De Mundo*, c. iv.

this early writer did not regard earthquakes as necessarily isolated events, but that some of them consisted of a succession of backward and forward vibratory motions. He also distinguishes between the total duration of an earthquake and the length of, and intervals between, a series of shocks. Aristotle had, in fact, some idea of what modern writers upon ordinary earthquakes would term 'modality.'

The earliest writer who had the idea that an earthquake was a pulse-like motion propagated through solid ground appears to have been Francisci Travagini, who, in 1679, wrote upon an earthquake which in 1667 had overthrown Ragusa. The method in which the pulses were propagated he illustrated by experiments.

Hooke, who, in 1690, delivered discourses on earthquakes before the Royal Society, divides these phenomena according to the geological effects they have produced; thus there is a *genus* producing elevations, a *genus* producing sinkings, a *genus* producing conversions and transportations, and a *genus* which produces what, in modern language, we should term metamorphic action.

Woodward, in his 'Natural History,' written in 1695, speaks of earthquakes as being agitations and concussions produced by water in the interior of the earth coming in contact with internal fires.

Stuckeley observed that an earthquake was 'a tremor of the earth,' to be explained as a vibration in a solid. The Rev. John Mitchell, writing in 1760, says that the motion of the earth in earthquakes is partly tremulous and partly propagated by waves.

From these few examples, to which might be added many more, it will be seen that an earthquake disturbance has usually been regarded as a concussion, vibration, trembling, or undulatory movement. Further, it can be seen in narratives of earthquakes that it had been often observed that these tremblings and shakings continued over a certain period of time. Although it had been noticed that large areas were almost simultaneously affected by these disturbances, no definite idea appears to have existed as to how earthquake motion was propagated. Usually it was assumed that the disturbance spread through subterranean channels.

The first true conception of earthquake motion and the manner of its propagation is due to Dr. Thomas Young, who suggested that earthquake motion was vibratory, and it might be 'propagated through the earth nearly in the same manner as a noise is conveyed through the air.' The same idea was moulded into a more definite form by Gay Lussac.

The first accurate definition of an earthquake is due to Mr. Robert Mallet, who, after collecting and examining many facts connected with earthquake phenomena, and reasoning on these, with the help of known laws connected with the production and propagation of waves of various descriptions, formulated his views as follows :—

An earthquake is 'the transit of a wave or waves of elastic compression in any direction from vertically upwards to horizontally, in any azimuth, through the crust and surface of the earth, from any centre of impulse or from more than one, and which may be attended with sound and tidal waves, dependent upon the impulse and upon circumstances of position as to sea and land.'

In brief, so far as motion in the earth is concerned, Mallet defined an earthquake as being a motion due to the transit of waves of elastic compression. In many cases it is possible that this is strictly true, but in succeeding pages it will be shown that earthquake motion may also be due to the transit of waves of elastic distortion.

To obtain a true idea of earthquake motion is a matter of cardinal importance, as it forms the key-stone of many investigations.

If we know the nature of the motion produced by an earthquake, we are aided in tracking it to its origin, and in reasoning as to how it was produced. If our knowledge of the nature of the motion of an earthquake is incorrect, it will be impossible for us intelligently to construct buildings to withstand the effects of these disturbances. We have thus to consider, in this portion of seismology, a point of great scientific importance, and shall deal with it at some length.

Nature of Elastic Waves and Vibrations.—When it is stated that an earthquake consists of elastic waves of compression and distortion, the student of physics has a clear idea of what is meant and a knowledge of the mechanical laws which govern such disturbances. The ordinary reader, however, and the majority of the inhabitants of earthquake countries, who of all people have the greatest interest in this matter, may not have so clear a conception, and it will, therefore, not be out of place to give some general explanation on this point.

The ordinary idea of a wave is that it is a disturbance similar to that which we often see in water. Waves like these must not, however, be confounded with elastic waves. A disturbance produced in water, say, for instance, by dropping a stone into a pond, is propagated outwards by the action of gravity. First, a ridge of water is raised up by the stone passing beneath the surface. As this ridge falls towards its normal position in virtue of its weight, it raises a second ridge. This second ridge raises a third ridge, and so on. The water moves vertically up and down, whilst the wave itself is propagated horizontally.

To understand what is meant by elastic waves, it is

first necessary to understand what is meant by the term elastic. In popular language the term elastic is confined to substances like india-rubber, and but seldom to rock-like materials, through which earthquake waves are propagated. India-rubber is called elastic because after we remove a compressive force it has a tendency to spring back to its original shape. The elastic force of the india-rubber is in this case the force which causes it to resist a change of form. Now, a piece of rock may, up to a certain point, like the india-rubber, be compressed, and when the compressing force is removed it will also tend to resume its original form. However, as the rock offers more resistance to the compressing force than the india-rubber offers, we say that it is the more elastic. It may be here observed that a substance like granite offers great resistance, not only to compression or a change of volume, but also to a change of form or shape; whereas a substance like air, which is also elastic, only offers resistance to compression, but not to a change of shape.

With these ideas before us we will now proceed to consider how, after a body has been suddenly compressed or distorted, this disturbance is propagated through the mass. For the elastic body let us take a long spiral spring hung from the ceiling of a room and kept slightly stretched by a weight. If we give this weight an upward tap from below, say with a hammer, we shall observe a pulse-like wave which runs up the spring until it reaches the ceiling of the room. Here it will, so to speak, rebound, like a billiard ball from the end of a table, and run towards the weight from which it started. Whilst this is going on we may also observe that the weight is moving up and down.

Here, then, we have two distinct things to observe--- one being the transmission of motion up to the ceiling,

which we may liken to the transmission of an earthquake
wave between two distant localities on the earth's surface,
and the other being the up and down motion of our
weight, which we may compare to the backward and forward swinging which we experience at the time of an
earthquake.

These two motions—namely, the pulse-like wave produced by the transmission of motion, and the backward
and forward oscillation of the weight or of any point on the
spring—must be carefully distinguished from each other.

First, we will consider the backward and forward motion
of the weight. The distance through which the weight
moves depends upon the force of the blow. The number
of up and down oscillations it makes, say in a second,
depends upon the stiffness of the spring. The weight,
supposing it to be always the same, will move more quickly
at the end of a stiff spring than at the end of a flaccid
one; that is to say, its velocity is quicker. As in any
given spring the number of up and down oscillations are
always the same in a given interval of time, if these
oscillations are of great extent, the weight must move
more quickly with large than with small oscillations.

At the time of an earthquake the manner in which
we are moved backward and forward is very similar to the
manner in which the weight is moved. If we stand on a
hard rock-like granite, we are to a great extent placed as
if we were attached to a stiff quickly-vibrating spring. If,
however, we are on a soft rock, it is more like being on a
loose flaccid spring.

All that has thus far been considered has been a backward and forward kind of motion, where there is a rectilinear *compression* and *extension* amongst the particles on
which we stand.

We might, however, imagine our rock, which for the

moment we will consider to be a square column, to be
twisted, and thus have its *shape* altered. When the
twisting force is taken off it seems evident that the
column would endeavour to untwist itself or regain its
original form. Now the force which a body offers against
a change of *volume* may be very different from that which
it offers against a change of *form*.

In disturbances which take place in the rocky crust
of our earth, it would seem possible that we may have
vibrations set up which are either compressions and ex-
tensions or twistings and distortions. These may take
place separately or simultaneously, or we may have resultant motions due to their combination.

The following are examples of possible causes which
might give rise to these different orders of disturbance:—

1. Imagine a large area stretched by elevation until
it reaches the limit of its elasticity and cracks. After
cracking, in consequence of its elasticity, it will fly back
over the whole area like a broken spring, and each point
in the area will oscillate round its new position of equilibrium. In this case there will be no waves of distortion
excepting near the end of the crack, where waves are
transmitted in a direction parallel to the fissure.

2. The ground is broken and slips either up, down,
or sideways, as we see to have taken place in the production of faults. Here we get distortion in the direction
of the movement, and waves are produced by the elastic
force of the rock, causing it to spring back from its distorted form. In a case like this the production of a
fissure running north and south might give rise to north
and south vibrations, which would be propagated end on
towards the north and south, but broadside on towards
the east and west. With disturbances of this kind, on
account of the want of homogeneousness in the materials

in which they are produced, we should expect to find waves of compression and extension.

3. A truly spherical cavity is suddenly formed by the explosion of steam in the midst of an elastic medium. In this case all the waves will be those of compression, each particle moving backward and forward along a radius.

Should the cavity, instead of being truly spherical, be irregular, it is evident that, in addition to the normal vibration of compression, transverse waves of distortion will be more or less pronounced, depending upon the nature of the cavity.

The combination of these two sets of vibrations may cause a point in the earth to move in a circle, an ellipse, the form of a figure eight, and in other curves similar to these, which are produced by apparatus designed to show the combination of harmonic motion. From these examples it will be seen that we have therefore to consider two kinds of vibrations—one produced by compression or the alteration of volume, and the other produced by an alteration in shape.

Now the resistance which a body offers, either to a change in its volume or in its shape, is called its elasticity, and the law which governs the backward and forward motion of a particle under the influence of this elasticity may be expressed as follows :

If T be the time of vibration, or the time taken by a particle to make one complete backward and forward swing, D the density of the material of which this particle forms a part, and E the proper modulus of elasticity of the material, then,

$$T = 2\,\pi\,\sqrt{\frac{D}{E}}$$

From this formula, $T = 2\,\pi\,\sqrt{\frac{D}{E}}$, we see that the

time of vibration of the earth during an earthquake, or
the rate at which we are shaken backwards and forwards,
varies directly as the square root of the density of the
material on which we stand, and inversely as the square
root of a number proportional to its elasticity.

Velocity and Acceleration of an Earth Particle.—
Another important point, which the practical seismologist
has often brought to his notice, is the question of the
velocity with which an earth particle moves.   According
to the formula, $T = 2 \pi \sqrt{\dfrac{D}{E}}$, we should expect that a
particle would make each semi-vibration in an equal time,
and from a knowledge of the density and elastic moduli
of a body this time might be calculated.   Although the
time of a semi-oscillation may be constant, we must bear
in mind that, like the bob of a pendulum during each of
its swings, the particle starts from rest, increases in velocity until it reaches the middle portion of its half swing,
from which it gradually decreases in speed until it reaches
zero, when it again commences a similar motion in the
opposite direction.

These pendulum-like vibrations are sometimes spoken
of as simple harmonic motions.   If we know the distance
through which an earthquake moves in making a single
swing, and the time taken in making this swing, on the
assumption that the motion is simple harmonic we can
easily calculate the maximum velocity with which the
particle moves.

Thus, if an earth particle takes one second to complete a semi-oscillation, half of which, or the amplitude
of the motion, equals a, the maximum velocity equals
$\pi \times a$.

Again, assuming the earth vibrations to be simple
harmonic, the maximum acceleration or rate of change in

E

velocity will come about at the ends of each semi-oscillation; and if v be the maximum velocity of the particle,
and a the amplitude or half semi-oscillation, then the
maximum acceleration equals $\dfrac{\text{v}^2}{a}$.

Later on it will be shown, as the result of experiment,
that certain of the more important earth oscillations in
an earthquake are not simple harmonic motion. Nevertheless the above remarks will be of assistance in showing how the velocity and other elements connected with
the motion of an earth particle, which are required by the
practical seismologist, may be calculated, irrespective of
assumptions as to the nature of the motion.

Propagation of a Disturbance.—We may next consider the manner in which a disturbance, in which there
are both vibrations of compression and of distortion, is
propagated. The first or normal set of vibrations are propagated in a manner similar to that in which sound
vibrations are propagated. From a centre of disturbance
these movements approach an observer at a distant
station, so to speak, end on. The other vibrations have
a direction of motion similar to that which we believe to
exist in a ray of light. These would approach the observer broadside on.

If the disturbance passed through a formation like a
series of perfectly laminated slates, each of these two sets
of vibration might be subdivided, and we should then
obtain what Mallet has termed ordinary and extraordinary
normal and transverse vibrations.

In consequence of the difference in the elastic forces
on which the propagation of these two kinds of vibration
depends, the normal vibrations are transmitted faster than
the transversal ones—that is to say, if an earthquake
originated from a blow, the first thing that would be

felt at a point distant from the origin of the shock would be a backward and forward motion in the direction to and from the origin, and then, a short interval afterwards, a motion transversal, or at right angles to this, would be experienced.

From the mathematical theory of vibratory motions it is possible to calculate the velocity with which a disturbance is propagated. As the result of these investigations it has been shown that normal vibrations travel more quickly than transverse vibrations.

Deductions from experiments on small specimens are, however, invalidated by the fact that the specimens used for experiments are, of course, nearly homogeneous, whilst the earthquake passes through a mass which is heterogeneous and more or less fissured.   Mallet, by experiments 'on the compressibility of solid cubes of these rocks, obtained the mean modulus of elasticity,' with the result that 'nearly seven-eighths of the full velocity of wave-transit due to the material, if solid and continuous, is lost by reason of the heterogeneity and discontinuity of the rocky masses as they are found piled together in nature.' The full velocities of wave-transit, as calculated by Mallet from a theorem given by Poisson, were—

For slate and quartz transverse to lamination, 9,691 feet per second.
„   „   in line of lamination,   5,415   „   „

This more rapid transmission in a direction transverse to the lamination, Mr. Mallet observes, may be more than counterbalanced by the discontinuity of the mass transverse to the same direction.

The Intensity of an Earthquake.—The intensity of an earthquake is best estimated by the intensity of the forces which are brought to bear on bodies placed on the earth's surface.   These forces are evidently proportional to the rate of change of velocity in the body, and, as the destructive effect will be proportional to the maximum forces, we may consistently indicate the intensity of an earthquake by giving the maximum acceleration to which bodies were subject during the disturbance. On the assumption that the motion of a point on the earth's surface is simple harmonic, the maximum acceleration is directly as the maximum velocity and inversely as the amplitude of motion, or as $\frac{v^2}{a}$ where v indicates velocity and a amplitude.

The next question of importance is to determine the manner in which earthquake energy becomes dissipated— that is, to compare together the intensity of an earthquake as recorded at two or more points at different distances from the origin. First let us imagine the origin of our earthquake to be surrounded by concentric shells, each of which is the breadth of the vibration of a particle. Going outwards from the centre, each successive shell will contain a greater number of particles, this number increasing directly as the square of the distance from the origin. Let the blow have its origin at the centre, and give a vibratory movement to the particles in one of the shells near the centre.

This shell may be supposed to possess a certain amount of energy, which will be measured by its mass and the square of the velocity of its particles. In transferring this energy to the neighbouring shell which surrounds it, because it has to set in motion a greater number of particles than it contains itself, the energy in any one particle of the second layer will be less than the energy in any one particle in the first layer ; the total energy in the second shell, however, will be equal to the total energy in the first shell. Neglecting the energy lost during the transfer, if the energy in a particle of the first shell at any particular phase of the motion be K_1, and the energy in a particle of the second shell K_2, these quantities are to each other inversely as the masses of the shells—that is, inversely as the squares of the mean radii of the shells.

In symbols, $$\frac{K_2}{K_1} = \frac{r_1^2}{r_2^2} \qquad \cdots \cdots \quad (1)$$

Assuming that energy is dissipated,

$$\frac{K_2}{K_1} > \frac{r_1^2}{r_2^2} = f\, \frac{r_1^2}{r_2^2} \qquad \cdots \cdots \quad (2)$$

where $f < 1$ is the rate of dissipation of energy which is assumed to be constant.

Area of greatest Overturning Moment.—Although the rate of dissipation of the impulsive effects of an earthquake may follow a law like that just enumerated, it must be remembered that if the depth of the origin is comparable with the radius of the area which is shaken, the maximum impulsive effect as exhibited by the actual destruction on the surface may not be immediately above the origin where buildings have simply been lifted vertically up and down, but at some distance from this point, where the impulsive effort has been more oblique.

At the *epicentrum* we have the maximum of the true intensity as measured by the acceleration of a particle, or the height to which a body might be projected, but it will be at some distance from this where we shall have the maximum intensity as exhibited by an overturning effort.

This will be rendered clear by the following diagram.

In the accompanying diagram let o be the origin of a shock, and o c the seismic vertical equal to r. Let the direct or normal shock emerge at c, c_1, c_2, and at the angles θ_1, θ_2, &c.

Assuming that the displacement of an earth particle at C equals C B, and at c_1 equals $c_1 b_1$, and at c_2 equals $c_2 b_2$, &c., and let these displacements C B, $c_1 b_1$, $c_2 b_2$, &c., for the sake of argument, vary inversely as r, r_1, r_2, &c.

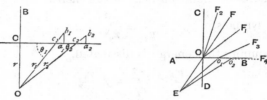

FIG. 9.

The question is to determine where the horizontal component C A of these normal motions is a maximum.

First observe that the triangle O C c is similar to a, b, c.

Also $r = \dfrac{h}{\sin \theta}$, and therefore the normal component $c_1 b_1$ at c_1 is equal to $c \dfrac{\sin \theta}{h}$.

Also $c_1 a_1 = c_1 b_1$, $\cos \theta$.

$$\therefore c_1 a = c \frac{\sin \theta \cos \theta}{h} = \frac{c}{h} \cdot \frac{\sin 2\theta}{2},$$

and $\sin 2\theta$ is greatest when $2\theta = 90°$ or $\theta = 45°$.

That is to say, the horizontal component reaches a maximum where the angle of emergence equals 45°.

This question has been discussed on the assumption that the amplitude of an earth particle varies inversely as its distance from the origin of the shock. Should we, however, assume that this amplitude varies inversely as the square of the distance from the origin, we are led to the result that the area of greatest disturbance is nearer to the point where the angle of emergence is 55° 44′ 9″.

Both of these methods are referred to by Mallet, but the first is considered as probably the more correct.

Earthquake Waves.—Hitherto we have chiefly considered earthquake vibrations; now we will say a few words about earthquake waves. If we strike a long iron rod at one end, we can imagine that, as in the long spring, a pulse-like motion is transmitted. If the rod be struck quickly, the pulses will rapidly succeed each other, and if struck slowly the pulses will be at longer intervals. Each individual pulse, however, will travel along the rod at the same rate, and hence the distance between any two will remain constant; but that distance will depend on the interval between the blows producing these pulses being equal to the distance travelled by one pulse before the next blow is struck.

From this we see that an irregular disturbance will produce an irregular succession of motions; some will be like long undulations in a wide deep ocean, whilst others will be like ripples in a shallow bay. Again, consider the bar to be struck one blow only, and then left to itself. The bar will propagate a series of pulses along its length, due to the out and in vibration of its end. These will succeed each other at regular intervals, and will be mixed up with the pulses we have previously considered.

From this we see that in an earthquake, if it be produced by one blow, the motion will be isochronous in its character; but if it be due to a succession of blows at regular intervals, the motion will be the resultant of a series of isochronous motions, and will be periodical. If the impulses are irregular, you have a motion which is the resultant of a number of isochronous motions due to each impulse, but these compounded together in a different manner at each instant during the earthquake, and giving as a result a motion which is in no sense isochron-

ous. This approaches more nearly to the actual motions
we feel as earthquakes.

If we can imagine the ground shaken by an earthquake, made of a transparent material which transmitted
less light when compressed, and we could look down upon
a long extent of this at the time of an earthquake, we
should see a series of dark bands indicating strips of
country which were compressed. The distances between
these bands might be irregular. Keeping our attention
on one particular band, this would be seen to travel
forward in a direction from the source. If we kept our
eye on one particular point, it would appear to open and
shut, becoming light and dark alternately.

As to the existence of these elastic waves in actual
earthquakes we have no direct experimental evidence.
The only kind of wave with which we are familiar is a true
surface undulation, which, although having the appearance
of a water-wave, may nevertheless represent a district of
compression.

CHAPTER IV.

EARTHQUAKE MOTION AS DEDUCED FROM EXPERIMENT.

Experiments with falling weights—Experiments with explosives—Results obtained from experiments—Relative motion of two adjacent points—The effect of hills and excavations upon the propagation of vibrations—The intensity of artificial disturbances—Velocity with which earth vibrations are propagated—Experiments of Mallet—Experiments of Abbot—Experiments in Japan—Mallet's results—Abbot's results—Results obtained in Japan.

Experiments with Falling Weights.—A series of experiments, as the nature of the disturbance produced in the surface of the earth when a heavy weight is allowed to fall on it, was begun in November 1880 by Mr. T. Gray and the author. These experiments were carried out at the Akabane Engineering Works in Tokio. The weight used was a ball of iron weighing about a ton, which in the different experiments was allowed to fall from heights varying between ten and thirty-five feet. The position of the place where the ball was allowed to fall was such that in one direction the vibrations were transmitted up the side of a steep hill, in another direction across a pond with perpendicular sides, and in another direction across a level plain the material of which consisted for the most part of hardened mud extending to a very considerable depth. The vibrations produced by the fall of the ball were transmitted through this hard mud with considerable intensity to a distance of between 300 and 400 feet.

The object of the experiment was to find the nature of the vibrations produced in the crust of the earth by such a blow, the velocity of transmission through this comparatively soft material, the effect of hills and excavations in cutting off such disturbances, and the law according to which the amplitude of the vibrations diminishes with the distance from the source.

A considerable variety of apparatus was used during these experiments, but the most reliable results were obtained from the records of a rolling sphere seismograph, which wrote the vibrations on a stationary plate, and from the records of two bracket seismographs, similar to Professor Ewing's horizontal lever seismographs, which gave a record of the vibrations as two rectangular components on a moving plate of smoked glass.

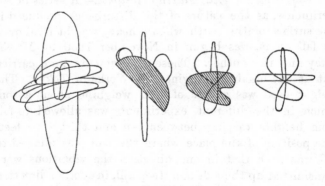

FIG. 10.

The general result as to the nature of the disturbance was that two distinct sets of vibrations were set up by the blow. In one set the direction of motion was along a line joining the point of observation with the point from which the disturbance emanated; in the other set the direction of motion was at right angles to that line. The nature of the resultant motion will be gathered from fig. 10, which is taken from the records drawn by the rolling sphere seismograph at a distance of 50 feet, 100 feet, and 200 feet respectively from the point where the ball struck the ground. The direct or normal vibrations reached the instrument first, and were followed at an interval depending on the distance of the instrument from the origin by the transverse vibrations. From the records of these two sets of vibrations as separated by the bracket seismographs, combined with the known rate of motion of the glass plate, the velocity of transmission was found to be, for normal vibrations 446–438 feet per second, and for transverse vibrations 357–353 feet per second.

The effect of the hill in cutting off the disturbance seemed to be slight, but the direction of the vibrations which ascended the side was mostly transverse. The pond, on the other hand, seemed completely to cut off the disturbance, which, however, gradually crept round the side, so that only a comparatively small triangular area was in shadow.

The amplitude of the vibrations diminished directly as the distance increased for some distance from the origin, but at greater distance the rate of diminution seemed to be slower. The transverse vibration seemed to die out less quickly than the normal vibrations.[1]

These experiments were afterwards very considerably extended by the author. In these later experiments charges of from one to two pounds of dynamite were placed in bore-holes of various depths and exploded by means of electricity. The results obtained confirmed the conclusions already arrived at from the former experiments. The experiments on velocity, however, seemed to indicate

[1] See *Phil. Trans. R. S.*, Part III. 1882.

that the higher the initial impulse the greater was the velocity. The velocity of propagation of the transverse vibrations seemed to approach more and more to that of the directed vibrations as the distance from the origin of disturbance increased. Fig. 11 shows the nature of the record obtained from the explosion of two pounds of dynamite at the bottom of a bore-hole eight feet deep.

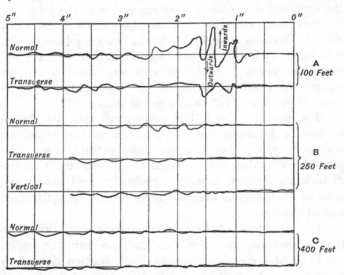

FIG. 11.—Records obtained at three stations of the motion of the ground produced by the explosion of 2 lbs. of dynamite.

These records show the interval of time which elapsed between the arrival of the normal and the transverse vibrations at points distant 100, 250, and 400 feet from the bore-hole. In the case of the 100-feet station it will be observed that the motion towards the origin is greater than that from the origin. It is also to be noticed that the period of vibration becomes greater as the distance from the origin increases.

The Intensity of Artificial Disturbances.—The data which we have at our disposal for determining the intensity of an earth particle which has been caused to vibrate by the explosion of a charge of dynamite are a series of records similar to that given on p. 60. These disturbances are practically surface movements, and may be compared with the movements of an earthquake which spreads over an area the radius of which is great as compared with its depth.

To find the mean acceleration of an earth particle, which quantity has been taken to represent intensity, during any simple backward or forward motion of the earth, it will be first necessary to determine the amplitude of this motion and its maximum velocity, the mean acceleration being equal to $\dfrac{v^2}{2\,A}$.

The second and third movements in a shock invariably exhibited the greatest intensity, and to a distance of 400 feet from the origin, where about three pounds of dynamite had been exploded in a bore-hole about six feet deep, these intensities decreased directly as the distance from the origin. The less intense movements also decreased directly as the distance from the origin to a certain point, but after that

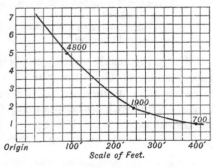

FIG. 12.

they decreased more slowly. A mean result of the more prominent vibrations in four sets of experiments is shown in the curve, fig. 12, where the horizontal measurements represent distance from the origin in feet, and the vertical measurements mean acceleration in thousands of milli-metres per second.

This curve approximates to an equi-angular hyperbola. The area between the curve and its asymptotes is propor-tional to the whole energy of the shock. The area of the diagram is proportional to the energy given up to the ground by the explosion of three pounds of dynamite. If we call the unit shock the effect produced by the explosion of one pound of dynamite, the above artificial earthquake had an intensity equal to three.

The only other investigations which have been made in this interesting branch of observational seismology are those by Mr. Robert Mallet,[1] and those by General Henry L. Abbot.[2]

Mallet's Results.—The velocity with which earth vibrations were transmitted as deduced by Mr. Mallet were as follows :—

	Feet per second
In sand	824·915
In contorted stratified rock, quartz, and slate at Holyhead	1,088·669
In discontinuous and much shattered granite	1,306·425
In more solid granite	1,664·574

A striking result which was obtained by Mallet in his experiments at Holyhead was that the transit velocity increases with an increase in the intensity of the initial shock. Thus with a charge of 12,000 pounds of powder the transit rate was 1,373 feet per second, whilst with 2,100 pounds the transit rate was 1,099 feet per second. In

---

[1] *Report of the British Association*, 1851.

[2] 'On the Velocity of Transmission of Earth Waves,' by General H. L. Abbot, *American Journal of Science and Arts*, vol. xv. March 1878; 'Shock of the Explosion at Hallet's Point,' by Bvt. Brig.-Gen. Henry L. Abbot, read before the Essayons Club of the Corps of Engineers, Nov. 1876.

these experiments tremors were observed as preceding and following the main shock.

Abbot's Results.—The important results obtained by General Abbot are contained in the following table :—

No. of Observation	Date	Cause of Shock	Distance to Station in miles	Type of Seismometer	Velocity in feet per second
1	Aug. 18, 1876	200 lbs. of dynamite	5 ±	B	5,280
2	Sept. 24, 1876	Hallet's Point Explosion	5·134	A	3,873
3	,,	,, ,, ,,	8·330	B	8,300
4	,,	,, ,, ,,	9·333	A	4,521
5	,,	,, ,, ,,	12·769	B	5,309
6	Oct. 10, 1876	70 lbs. dynamite	1·360	A	1,240
7	Sept. 6, 1877	400 ,, ,,	1·169	A	3,428
8	,,	,, ,, ,,	1·169	B	8,814
9	Sept. 12, 1877	200 ,, ,,	1·340	·A ´	6,730
10	,,	,, ,, ,,	1·340	B	8,730
11	,,	70 ,, ,,	1·340	A	5,559
12	,, .	,, ,, ,,	1·340	B	8,415

A seismometer of type A means that the telescope used in observing the tremor produced on the surface of a vessel of mercury by the passage of the shock had a magnification of 6, whilst a telescope of the type B had a magnification of 12.

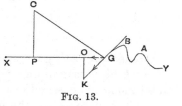

FIG. 13.

The mean velocity given by six observations with type A is 4,225 feet per second, while that given by the same number with type B is 7,475 feet per second.

If we assume that the first tremor observed in the mercury is to determine the true rate of transmission, General Abbot tells us that we must reject all observations made with type A, inasmuch as they do not reveal the velocity of the leading tremor. However, he also tells us that a still higher power above 12 might have detected still earlier tremors.

When gunpowder was the explosive, the observers noted that the disturbance observed in the mercury took a much longer time to reach a maximum than it did when dynamite was employed.

It was also observed that explosions fired beneath deep water gave a higher velocity than similar explosions which took place beneath shallow water. In the latter case much of the energy was probably expended in throwing a jet of water into the air.

Another point which was observed appears to have been that the rate varied with the initial shock. Thus :—

	Feet per second
400 lbs. of dynamite gave	8,814
200 „ „	8,730
70 „ powder (deep) gave	8,415

Also it is probable that the rate of a wave diminished with its advance. For,

	Feet per second
200 lbs. of dynamite gave for 1 mile . .	8,730
„ „ „ „ 5 miles . .	5,250
50,000 „ „ „ 8 „ . .	8,300
„ „ „ „ 13½ „ . .	5,300

General Abbot's general conclusions are :—

1. A high magnifying power of telescope is essential in seismometric observations.

2. The more violent the initial shock the higher is the velocity of transmission.

3. This velocity diminishes as the general wave advances.

4. The movements of the earth's crust are complex, consisting of many short waves first, increasing and then decreasing in amplitude; and with a detonating explosive the interval between the first wave and the maximum

wave, at any station, is shorter than with a slow burning explosive.

Results obtained in Jápan.—From some experiments made by the author in the grounds of the Meteorological Department in Tokio, the following results were obtained:—

No. of Explosion		Velocity in feet per second for the first 200 ft. (A to B)	Velocity in feet per second for the second 200 ft. (B to C)	Velocity in feet per second for 400 ft. (A to C)	Number of Cartridges of Dynamite (6 = 1 lb.)
Vertical vibrations	I.	464	186	265	8·3
	III.	—	211	—	10·1
	IV.	352	234	281	7·1
	V.	343	232	277	5·0
Normal vibrations	VI.	—	—	407	10·0
	VII.	—	—	516	12·5
Transverse vibrations	VIII.	—	—	344	12·5

The general results to be deduced from the above appear to be:—

1. For vertical motion.
    (*a*) For the first 200 feet. The velocity depends upon the initial force—the greater the charge of dynamite the greater the velocity.
    (*b*) For the second 200 feet. The above law only appears in experiments IV. and V., but it must be remembered that the origins of I. and III. were farther removed from A than IV. and V.
    The speed of the wave during the second 200 feet is always less than during the first 200 feet.
2. For normal vibrations.
    Here the speed between A and C is all that was measured, but we again see that the greater the initial force, or the nearer we are to the origin of the disturbance, the greater is the velocity. This velocity is greater than the velocity of the vertical or transverse vibrations

F

3. For transverse vibrations.

   If we assume that the vertical vibrations are a
   component of the transverse motions we see the
   same law as before—namely, that the nearer we
   are to the origin of the disturbance the greater
   is the speed with which that disturbance is propagated.

It will be observed that the chief law here enunciated
respecting the decrease in speed of earth vibrations is the
same as that pointed out by General Abbot, from which it
only differs by its being in all cases proved without the
introduction of personal errors, for the same explosion,
along the same line of ground and for different kinds of
vibrations.

CHAPTER V.

EARTHQUAKE MOTION AS DEDUCED FROM OBSERVATION ON EARTHQUAKES.

Result of feelings—The direction of motion—Instruments as indicators of direction—Duration of an earthquake—Period of vibration—The amplitude of earth movements—Side of greatest motion—Intensity of earthquakes—Velocity and acceleration of an earth particle—Absolute intensity of an earthquake—Radiation of an earthquake—Velocity of propagation.

Result of Feelings.—As the result of our experiences, and by observations upon the movements produced in various bodies, we can say that an ordinary earthquake consists of a number of backward and forward motions of the ground following each other in quick succession. Sometimes these commence and die out so gently that those who have endeavoured to time the duration of an earthquake have found it difficult to say when the shock commenced and when it ended. This was a difficulty which Mr. James Bissett in Yokohama, and the author in Tokio, had to contend against when, in 1878, they commenced to time shocks between these two places.

Sometimes these motions gradually increase to a maximum and then die out as gradually as they commenced.

Sometimes the maximum comes suddenly, and at other times during an earthquake our feelings distinctly tell us that there are several maxima.

These have been the experiences of many observers, and have been recorded by writers since the earliest times. Mallet devotes a chapter to a consideration of the tremulous motion that precedes and follows a shock, and he tells us that a single shock is an absolute impossibility. In speaking of earthquakes, he says: ' The almost universal succession of phenomena ·recorded in earthquakes is, first a trembling, then a severe shock, or several in quick succession, and then a trembling gradually but rapidly becoming insensible.'

A quantitative and exact knowledge of the nature of earthquake motion has only been attained of late years. The chief results which investigators have aimed at have been the measurement of the amplitude, the period, the direction, and the duration of the motions which constitute an earthquake. Attention has also been given to the velocity with which a disturbance is propagated.

The Direction of Motion.—One of the most ordinary observations which are made about an earthquake is its direction. If we were to ask the inhabitants of a town which had been shaken by an earthquake the direction of the motion they experienced, it is not unlikely that their replies would include all the points of the compass. Many, in consequence of their alarm, have not been able to make accurate observations. Others have been deceived by the motion of the building in which they were situated. Some tell us that the motion had been north and south, whilst others say that it was east and west. A certain number have recognised several motions, and amongst the rest there will be a few who have felt a wriggling or twisting. Leaving out exceptional cases, the general result obtained from personal observation as to the direction of an earthquake of moderate intensity is extremely indefinite, and the only satisfactory information to be obtained is that derived from instruments or from the effects of the earthquake exhibited in shattered buildings and bodies which had been overturned or projected.

By the direction in which walls, columns, and other objects had been overthrown or fractured, Mallet was enabled to determine the position of the origin of the Neapolitan earthquake. Similar phenomena have many times been taken advantage of by other investigators of earthquake phenomena. Effects produced upon structures are, however, only to be observed as the results of a destructive earthquake, at which time cities may be regarded as collections of seismometers. (*See* chapter on Effects in Buildings.)

To determine the direction of movement during a small earthquake, the most satisfactory method appears to be an appeal to instruments.

Instruments as Indicators of Direction.—The relative values of different kinds of instruments, such as columns, pendulums, and the like, as indicators of direction have already been discussed.

By the use of pendulum seismographs it has been shown that during an earthquake the ground may move in one, two, or several directions (see p. 21); and it is, generally speaking, only in those cases where we experience a decided shock in the disturbance that we can determine with any confidence the direction in which the motion has been propagated. Such directions are usually indicated by the major axis of certain more or less elliptical figures which have been drawn, which in themselves appear to indicate the combination of two rectilinear movements.

Results similar to those indicated by the records of pendulum seismographs have also been obtained upon moving plates with a double bracket seismograph. Thus,

in the earthquake which shook Tokio at 6 A.M. on July 5,
1881, there were indications of the following motions : —

Near the commencement of the shock the motion was
N. 112° E.   One and a half second after this, the direction of motion appears to have been N. 50° E.   In three-
fourths of a second more it gradually changed to a
direction N. 145° E., and after a similar interval to N.
62° E.   Half a second after this it was N. 132° E., and
four seconds later the motion was again in the original
direction—namely, N. 112° E.

These particular directions of motion have been
selected because they were so definitely indicated.

The commonest type of earthquake which is experienced in Japan, and probably also in other earthquake-
shaken districts, is the compound or diastrophic form.

That earthquakes often have motions compounded of
two sets of vibrations, has also been proved by the analysis of the records obtained from two component seismographs.   From an analysis of a record of this description,
Professor Ewing has shown that in the earthquake felt in
Tokio on March 11, 1881, there were approximate circular
(somewhat spiral) movements.

This leads us to the consideration of the twisting and
wriggling motions which are said to be experienced by
some observers.   Motions like these, which by the Italians
and Mexicans are called *vorticosi*, are usually supposed to
be the cause of objects like chimneys and gravestones
being rotated.   These phenomena, it will be seen from
what is said in the chapter upon the effects produced in
buildings, can be more easily explained upon the supposition of a simple rectilinear movement.

That at the time of an earthquake there may be
motion in more than one direction has been recognised
since the time of Aristotle ; and it is possible that two sets

of rectilinear motion, as, for instance, the normal and
transverse movements, may have led observers to imagine
that there has been a twisting motion taking place, and
this especially when the two sets of movements have
quickly succeeded each other.

Persons inside flexible buildings may possibly have
experienced more or less of a rotatory motion, although
the shock was rectilinear; the building assuming such a
motion in consequence of its construction and its position
with regard to the direction of the shock.

In the case of destructive earthquakes, especially at
points situated practically above the origin, the universal
testimony, Mallet tells us, is that a twisting, wriggling
motion in different planes, attended by an up-and-down
movement of greater range, is experienced. To such disturbances the word *sussultatore* is sometimes applied.
Mallet has given many elliptical and other closed curves
to illustrate the nature of such motions.

Duration of an Earthquake.—When reading accounts
of earthquakes it is often difficult to determine the length
of time a shaking was continuous. In Japan, in A.D. 745,
there was a shaking which is said to have lasted sixty
hours; and in A.D. 977 there were a series of shakings
lasting 300 days. Often we meet with records of disturbances which have lasted from twenty to seventy days.

At San Salvador, in 1879, more than 600 shocks were
felt within ten days; in 1850, at Honduras, there were
108 shocks in a week; in 1746, at Lima, 200 shocks were
felt in twenty-four hours; at the island of St. Thomas, in
1868, 283 shocks were felt during about ten hours.

Disturbances like these, which succeed each other with
sufficient rapidity to cause an almost continual trembling
in the ground, may be regarded as collectively forming
one great seismic effort which may last a minute, an hour,

a day, a week, or even several years. Strictly speaking, they are a series of separate earthquakes, the resultant vibrations of which more or less overlap. Whenever a large earthquake occurs it is generally succeeded by a large number of smaller shocks.

The seismic disturbance as regards time is, as Mallet remarks, very often 'like an occasional cannonade during a continuous but irregular rattle of musketry.' In the New Zealand earthquake of 1848, shocks continued for nearly five weeks, and during a large portion of the time there were at least 1,000 shocks per day.[1]

The earthquake of Lisbon, which in five minutes destroyed the whole town, was followed by a series of disturbances lasting over several months. After Basle had, on October 18, 1356, been laid in ruins, it is stated shocks followed each other for a period of a year. The Calabrian earthquake was continued with considerable strength for a year, and it is said that the earth did not come completely to rest for ten years. During this cannonade the heavy shocks announced, as they do in most earthquake countries at the present day, a series of weaker disturbances. In certain exceptional cases this order of events has been inverted, and slight shocks have announced the coming of heavy ones. Fuchs gives an example of this in the earthquake of Broussa, when the first shock was on February 28, 1855. On March 9 and 23 there were heavier shocks, but the heaviest did not arrive until March 28.

Under certain conditions it is possible to have a sensible vibration produced in the ground which is practically of unlimited duration; thus, for instance, it has been noticed that the falling of water at certain large waterfalls, by its continuous rhythmical impact on the

[1] *West. Rev.*, July 1849.

rocks, produces in them tremors which are to be observed at great distances. Of this the author convinced himself at the Falls of Niagara, where he observed the reflected and ever-moving image of the sun in a pool of water. Under favourable circumstances almost continual condensation of steam might take place in volcanic foci, each condensation giving rise to a blow sufficiently powerful to produce vibrations in the surrounding ground. Those who have stood near a large geyser, like the one in Iceland, when it makes an ineffectual effort to erupt, will recognise how powerful such a cause might be. Humboldt has remarked shocks on Vesuvius and Pichincha which were periodic, occurring twenty to thirty seconds before each ejection of vapour and ashes.

Earthquakes like these may be of vast extent, gradually spreading further and further outwards. This spreading of earth vibrations may be observed at a large factory containing heavy machinery or a steam hammer. After the machinery comes to rest, it is probably some time before the ground returns to rest. Examples of disturbances of this nature are spoken of under the head of Earth Tremors.

The record of the duration of an ordinary earthquake as observed at a given point is dependent upon the sensibility of our instruments.

Continuous motions perceptible to our senses without the aid of instruments usually last from thirty seconds to about two or three minutes. In Japan the shocks, as timed by watches, usually last from twenty to forty seconds. Occasionally a continuous shaking is felt for more than one and a half minutes, and cases have been recorded where the motion has continued for as much as four minutes and thirty-three seconds.

Seismometers having a multiplication of 6 to 12

usually indicate that motion continues longer than is perceptible to the senses.

Period of Vibration.—When an earthquake contains several prominent vibrations which might be called the *shocks* of the disturbance, our feelings tell us that these have occurred at unequal intervals.

About the time which is taken for the complete backward and forward oscillation of the ground which constitutes the shock a little has already been said. This was deduced from the records of disturbances as drawn by seismographs. From the same sources we can readily obtain the period of all the prominent vibrations in a disturbance.

In any given earthquake there are irregularities in period, and different earthquakes differ from each other. About the early attempts to determine the period of earth vibrations something has been said in the chapter on Earthquake Instruments.

In the earthquake of March 11 (referred to on p. 70) we find that both components commenced with a series of small vibrations, about five or six to the second; next came the shock, consisting of two complete vibrations executed in two seconds. In this it is to be observed that the motion eastwards was performed much more quickly than the motion westwards. Next, by reference to the east and west component, it is seen that there are a number of large vibrations, about one per second, on which a number of smaller motions are superposed. As the motion proceeds, these become less and less definitely pronounced and more irregular in their intervals, until finally the motion dies away.

This earthquake, as recorded at the author's house in Tokio, lasted about one and a half minute.

The same earthquake, as recorded by Professor Ewing

at a station situated about one and a half mile distant, but on flat ground, appears to have lasted four and a half minutes. The largest wave had a period of 0·7 second.

In the earthquake of March 8, 1881, there were on an average 1·4 vibrations per second. These vibrations were executed in a direction transverse to the line joining the observing station and the locality from which the disturbance must have originated as determined by time observations. It can, therefore, be assumed that these vibrations, having so slow a period, were transverse motions, this slowness or sluggishness being due to the fact that the modulus for distortion is less than the modulus which governs the propagation of normal vibrations.

The Amplitude of Earth Movements.—In making estimates of the distances through which we are moved backward and forward at the time of an earthquake, if we judge by our feelings, we may often be misled. If a person is out of doors and walking, an earthquake may take place sufficiently strong to cause chimneys to fall and unroof houses, which, so far as the actual shaking of the ground is concerned, will be passed by unnoticed. On the other hand, to persons indoors, especially on an upper story, it is impossible even for a tremor to pass by without creating considerable alarm by the angular movement that has been taken up by the building.

Many observers have endeavoured to make actual measurements of the maximum extent through which the earth moves at the time of an earthquake. Among the reports of the British Association for 1841 is the report of a committee which had been appointed ' for obtaining instruments and registers to record shocks of earthquakes in Scotland and Ireland. We read that in one earthquake which had been measured the displacement of the ground had been half an inch, and in another it had been less than half an inch. The instruments used to make these observations depended upon the inertia of pendulums which at the time of the disturbance were supposed to remain at rest. Observations similar to these have been made in Japan. One long series were made by Mr. E. Knipping for Dr. G. Wagener. They extended from November 1878 to April 1880, and were as follows :—

Number of Earthquakes	Maximum horizontal motion of the ground
10	·0 to 0·15 mm.
7	·15 „ 0·5 „
8	·5 „ 2·5 „
2	2·5 „ more „

With his apparatus for vertical motion Dr. Wagener also made observations on the absolute vertical motion. This seldom reached ·02 mm. The greatest value was that observed for the destructive shock of Feb. 22, 1880, which was ·56 mm.

By means of a number of instruments distributed at various localities round Tokio, the chief of which were pendulums with friction pointers to render them ' dead beat,' and with magnifying apparatus to show the actual motion of the ground, the author arrived at results similar to those obtained by Dr. Wagener—namely, that the earth's maximum horizontal motion at the time of a small earthquake was usually only the fraction of a millimetre, and it seldom exceeded three or four millimetres. When we get a motion of five or six millimetres, we usually find that brick and stone chimneys have been shattered.

The results obtained for vertical motion were also very small. In Tokio it is seldom that vertical motion can be detected, and when it is recorded it is seldom more than a millimetre.

These results, which were put forward some years ago, have since received confirmation by the use of a variety of instruments in the hands of different observers.

Mallet, in his account of the Neapolitan earthquake of 1857, approximated to the amplitude of an earth particle by observing the width, at the level of the centre of gravitÿ, of fissures formed through and remaining in great masses of very inelastic masonry.

Taking stations situated on or very nearly on the same line passing through the seismic vertical (*epicentrum*), Mallet observed the amplitude increased as some function of the distance, as will be seen from the following table :—

Station	Polla	La Sala	Certosa	Tramutola	Sarconi
Distance from Seismic Vertical in geographical miles . .	3·45	11·60	16·50	20·60	26·7
Amplitude in inches .	2·5	3·5	4·0	4·5	4·75

The possibility of a law such as this having an existence for places at a distance from the seismic vertical comparable with the vertical depth of the centrum will be shown farther on.

With regard to the maximum displacement of an earth particle, Mallet was of opinion that there was evidence to show that it had in some cases been over one foot. M. Abella, in an earthquake which occurred in the Philippines in 1881, made a rough observation of the motion of the earth to a distance of about *two metres*. This, as might be expected, was beyond the elastic limits of the material, and caused fissures to be formed, which were seen to open and shut.

Intensity of Earthquakes.—In speaking of the strength

of an earthquake, we usually employ terms like 'weak,'
' strong,' 'violent,' &c. Although these expressions, accompanied by illustration of the effects which an earthquake has produced, convey a general idea of the strength
of a shock as felt at some particular locality, our ideas are
nevertheless wanting in definiteness; and if we endeavour
to compare one shock with another, as a whole, our want
of exactness is augmented. We have seen that Palmieri's
seismograph indicates intensity by a certain number of
degrees, which, to a certain extent, is a measure of the
violence of the motion as indicated at a particular locality.
The degrees, as before stated, refer to the height to which,
in consequence of the shaking, a certain quantity of
mercury was washed in a tube, which is a function of the
depth of mercury in the tube, and also of the duration of
the disturbance.

From this it seems possible that a very slow motion
of small amplitude, continuing over a sufficient period of
time, might, if it agreed with the period of the mercury,
indicate an earthquake of many degrees of intensity,
whilst residents in the neighbourhood might not have
noticed the disturbance ; and, on the other hand, a short
but intense shock creating considerable destruction might
have been recorded as of only a few degrees of intensity.

Although objections like these might be raised to such
a method of recording intensity, in practice it would
appear that such results are not pronounced, and the
indications of the instrument usually give us approximate
indications of relative intensity.

In writing about the Neapolitan earthquake of 1857,
Mallet says that ' area alone affords no test of seismic
energy.'

The area over which a shock is felt will depend not
only upon the initial force of the disturbance, but also

upon the focal depth of a shock, the form and position of that focus, the duration of the disturbance, and the nature and arrangement of the materials which are shaken.

From observations in Japan, it is clearly shown that massive mountain ranges exert a considerable influence upon the extension of seismal disturbances. On one side of a large range of mountains large cities might be laid in ruins, whilst on the other side the disturbance creating this destruction might not be noticed.

Velocity and Acceleration of an Earth Particle.— We now pass on to methods of determining the intensity of an earthquake which are less arbitrary than those which have just been discussed. These methods have already been discussed when speaking of artificial disturbances, where it was shown that the intensity of an earthquake as measured by its destructive effects greatly depended upon the suddenness with which the backward and forward motions of the ground were commenced or ended.

Amongst the earlier investigators of seismic phenomena who observed that there existed a connection between the distance to which bodies had been projected during an earthquake and the suddenness or initial velocity with which the ground had been moved beneath them, was Professor Wenthrop of Cambridge, Massachusetts, who noted that bricks from his chimneys had, by the New England earthquake of 1755, been thrown thirty feet. From this and the known height of the chimney, he calculates that the bricks had been projected with an initial velocity of twenty-one feet per second.[1]

The calculations made by Mallet respecting the maximum velocity of an earth particle at the time of the Neapolitan earthquake in 1857 depended upon the overthrow, projection, and fracture of bodies.

[1] *Phil. Trans.,* L., 1755.

The principles which guided him in making these calculations will be understood from the following illustration.

If a column, A B C D, receive a shock or be suddenly

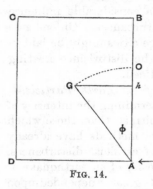

FIG. 14.

moved in the direction of the arrow, the centre of gravity, G, of this column will revolve round the edge, and tend to describe the path G O. If it passes O, the column will fall. The work done in such a case as this is equal to lifting the column through the height o h.

If G A $= a$, the angle G A $h = \phi$, and the weight of the body $= w$, then the above work equals

$$w a \ (1 - \cos \phi).$$

This must equal the work acquired—that is to say, the kinetic energy of rotation of the body, or

$$w a \ (1 - \cos \phi) = \frac{w w^2 \kappa^2}{2 \ g}.$$

Where w is the angular velocity of the body at starting, κ the radius of gyration round A, and g the velocity acquired by a falling body in one second. Whence

$$w^2 \ \kappa^2 = 2 \ g a \ (1 - \cos \phi),$$

but w, the angular velocity, is equal to the statical couple applied, divided by the moment of inertia, or,

$$w = \frac{v a \cos \phi}{\kappa^2},$$

squaring and substituting

$$v^2 = 2 g \times \frac{\kappa^2}{a} \times \frac{1 - \cos \phi}{\cos^2 \phi},$$

and since the length of the corresponding pendulum is
$$l = \frac{\kappa^2}{a},$$

$$v^2 = 2\,gl \times \frac{1 - \cos\,\phi}{\cos^2\,\phi}.$$

To apply this to any given case we must find the value of l or of $\frac{\kappa^2}{a}$.

Mallet finds these values for the cube, solid and hollow rectangular parallelopipeds, solid and hollow cylinders, &c. In these formulæ we have a direct connection between the dimensions and form of a body and the velocity with which the ground must move beneath it to cause its overthrow.

Not only is the case discussed for horizontal forces, but also for forces acting obliquely. Similar reasonings are applied to the productions of fractures in walls, but as there is uncertainty in our knowledge of the coefficient of force necessary to produce fracture. *through joints across* beds of masonry, the deductions ought not to be applied as the measures of velocity. Where the fractures occur at the base or in horizontal planes, or in those of the continuous beds of the masonry, or through homogeneous bodies, the uncertainty is not so great, and for cases like these Mallet gives several illustrations. The distance to which bodies had been projected,

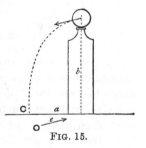

FIG. 15.

as, for example, ornaments from the tops of pedestals, coping-stones from the edges of roofs, were also used as means of determining the angle at which the shock had

G

emerged, or, if this be known, for determining the velocity.

Thus by a shock in the direction o c, a ball, A, on the top of a pedestal would describe a trajectory to the point c. Let the angle which o c makes with the horizon be e, the vertical height through which the ball has fallen be b, and the horizontal distance of projection be a; then

$$b = a \tan e + \frac{a^2}{4 \text{ H} \cos^2 e},$$

H being the height due to the velocity of projection. whence

$$\text{Tan } e = \frac{2 \text{ H} \pm \sqrt{4 \text{ H} (\text{H} + b) - a^2}}{a}.$$

$$\text{v}^2 = \frac{a^2 \, g}{2 \, \cos^2 e \, (b - a \tan e)}.$$

For the back motion or subnormal wave in the direction c o,

$$\text{Tan } e = \frac{2 \text{ H} \pm \sqrt{4 \text{ H} (\text{H} + b) - a^2}}{a}$$

$$\text{v}^2 = \frac{a^2 \, g}{2 \, \cos^2 e \, (b + a \tan e)}.$$

A serious error which may enter into calculations of this description when practically applied has been pointed out when speaking of columns as seismometers. It was then shown that such bodies before being overthrown may often be caused to rock, and therefore that their final overthrow may not have any direct connection with the impulse of the succeeding shock.

Another point to which attention must be drawn respecting the above calculations is that if there was no friction or adherence between the projected body and its pedestal, in consequence of its inertia it would be left behind by the forward motion of the shock, and simply drop at the foot of its support. In the case of frictional adherence it would be carried forward by the velocity acquired before this adherence was broken, and thrown in a direction *opposite* to that given in the figure—that is to say, in the direction of the shock.[1]

The Absolute Intensity of the Force exerted by an Earthquake.—No doubt it has occurred to many who have experienced an earthquake that the power which gave birth to such a disturbance must have been enormously great. The estimates which we shall make of the absolute amount of energy represented by an earthquake cannot, on account of the nature of the factors with which we deal, be regarded as accurate. They may, however, be of assistance in forming estimates of quantities about which we have at present no conception. One method of obtaining the result we seek is that which was employed by Mallet in his calculations respecting the Neapolitan earthquake. Although disbelieving in the general increment of temperature as we descend in the earth at an average rate of 1° F. for every fifty or sixty feet of descent, for want of better means, Mallet assumes this law to be true, and, knowing from a variety of observations the depth of various parts of the cavity from which the disturbance sprang, he calculates the temperatures of this cavity in various parts as due to its depth beneath the surface. Next, it is assumed that steam was suddenly admitted into this cavity, which might exert the greatest possible pressure due to the maximum temperature. This was calculated as being about 684 atmospheres.

[1] The solution is taken from Mallet's *Account of the Neapolitan Earthquake*, vol. i. p. 155.

Next, he determined the column of limestone necessary to balance such a pressure, which is about 8,550 feet in height. As the least thickness of strata above this cavity was 16,700 feet, the pressure of 684 atmospheres was not sufficient to blow away its cover, but if suddenly admitted or generated in the cavity it might have produced the wave of impulse by the sudden compression of the walls of the cavity.

The pressure of 684 atmospheres is equivalent to about 4·58 tons on the square inch, and, as the total area of the walls of the cavity is calculated at twenty-seven square miles, the total accumulated pressure would be more than 640,528 millions of tons. Mallet, however, shows that it is probable that the temperature of the focal cavity was much greater than that due to the hypogeal increment, and that therefore the pressure may have been greater.

The capability of producing the earthquake impulse, however, depends on the *suddenness* with which the steam is flashed off. According to the experiments of Boutigny and others, Mallet tells us that the most sudden production of steam would take place at a temperature of 500°–550° C., which is but a few degrees below that calculated for the mean focal depth.

Assuming the above calculated pressure to be true, and knowing the co-efficient of compression of the materials on which it acted, the volume of the wave at a given moment near the instant of starting—that is, at the focus—can be calculated, and from this the wave amplitude on reaching the surface may be deduced.

Proceeding backwards, if we have observed the wave amplitude, calculated the depth of the focus, and know the co-efficient of expansion, then the total compression may be calculated and the temperature due to the pressure producing this may be arrived at. In this way earthquakes may be used as a means of calculating subterranean temperature at depths that can never be attained experimentally.

A method of proceeding which is probably more definite than that adopted by Mallet would be the application of the method indicated when speaking of the intensity of artificial disturbances.

If for a given earthquake the origin of which is known we have determined by seismographs the mean acceleration of an earth particle at two or more stations at different distances from that origin, we are enabled to construct a curve of intensity the area between which and its asymptotes was shown to be a measure of the total intensity of the shock. Comparing this area with that of a unit disturbance produced, say, by the explosion of a pound of dynamite, one may approximately calculate in terms of this unit the initial intensity of the earthquake.

Radiation of an Earthquake.—The tremors preceding the more violent movements of an earthquake may be due, as Mallet has suggested,[1] to the free surface waves reaching a distant point before the direct vibrations.

The fact that earth vibrations produced by striking a blow on or near the surface of the ground are wholly obliterated in reaching a cutting or valley, there being no underground waves of distortion to crop up on the opposite side of the valley, indicates that the disturbance is one that travels on the surface; the same fact is illustrated when we endeavour to transmit vibrations through the side of a hill into a tunnel.

In the tunnel, although the distance may be small,

[1] *Neapolitan Earthquake,* ii. p. 300.

no sensible effects are produced, whilst the same disturbance may be recorded at a long distance from its origin
on the surface of the ground outside the tunnel.

Lastly, we may refer to the experiences of miners
underground.

Occasionally it has happened that miners when deep
underground, as in the Marienberg in the Saxon Erzge-
birge, have felt shocks which have not been noticed on the
surface.   These observations are rare, and it is possible
that they may be explained by the caving in of subterranean excavations.

The usual experience is, that if a shock is felt
underground it is also felt on the surface, as for example
in the lead mines in Derbyshire at the time of the
Lisbon disturbance (1755).

The most frequent observation, however, is that a
shock may be felt on the surface while it is not remarked
by the miners beneath the surface, as at Fahlun and
Presburg in November, 1823.

At the Comstock Lode in Colorado about twelve
years ago many earthquakes were felt.   On one particular day twenty-four were counted.   Superintendent
Charles Foreman told the author when he visited
Virginia City in 1882, that special observations were
made to determine whether these shocks were felt
as severely deep down in the mines as on the surface,
where they were on the verge of being destructive.
The universal testimony of many observers was that in
most cases they were not felt at all underground, and
when a shock was felt it was extremely feeble.   At
Takashima Colliery, in Japan, it is seldom that shocks
are felt underground.

The explanation of these latter observations appears
to be either that, in consequence of a smaller amplitude

of motion in the solid rocks beneath the surface as compared with the extent of motion on the surface, the disturbances are passed by unnoticed, or else the disturbance is, at a distance from its origin, practically confined to the surface.

Velocity of Propagation of an Earthquake.—Although many have written upon earthquakes and have endeavoured to give to us the velocity with which they were propagated, the subject is one about which we have as yet but little exact information.

The importance of this branch of investigation is undoubtedly great. By knowing the velocity with which an earthquake has travelled in various directions we are assisted in determining the locality of its origin ; we may possibly make important deductions respecting the nature of the medium through which it has passed ; perhaps also we may learn something regarding the intensity of the disturbance which created the earthquake. In the Report of the British Association for 1851 Mallet gives the table on next page, in which are placed together the approximate rates of transit of shocks of several earthquakes which he discusses. Some of these, it will be observed, are records of disturbances which must have passed through or across the bed of the ocean.

In Mallet's British Association Report for 1858, he gives data compiled by Mr. David Milne [1] respecting the Lisbon earthquakes of 1755 and 1761, from which data the tables of velocities (p. 89) have been calculated, omitting those which Mr. Mallet has marked as uncertain.

The distances are marked in degrees of seventy

[1] See *Edinburgh Phil. Trans.*, vol. xxxi.

Occasion and Place	Approx. rate in feet per second	Formation constituting Range on surface as far as known or conjectured	Authority
Rev. John Mitchell's guesses from the Lisbon earthquakes	1,760	Sea bottom, probably on slates, secondary and crystalline rocks	Mitchell
Von Humboldt's guesses from South America	1,760 to 2,464	From observations in various South American rocks in great part volcanic	Humboldt
Lisbon Earthquake of 1761.			
Lisbon to Corunna	1,994	Transition, carboniferous and granitoid	'Annual Register'
Lisbon to Cork	5,228	Transition, carboniferous crystalline slates and granitoid, probably, under sea bottom	,,
Lisbon to Santa Cruz	3,261	The same with many alterations	,,
Antilles.			
Pointe à Pitre to Cayenne (doubtful)	6,586	Probably volcanic rocks under sea bottom	Stier and Perrey's memorandum, Dijon
India.			
Cutch to Calcutta, 1819	1,173	Alluvial, secondary, granitoid and later igneous rocks	'Royal Asiatic Journal'
India, Nepauls, and basin of the Ganges, 1834:—			
Rungpur to Arrah	2,314	Deep alluvia, with occasional transition, carboniferous, granitoid, and later igneous rocks	'Royal Asiatic Journal'
Monghyr to Gorackpur	3,520		
Rungpur to Monghyr	990		
Rungpur to Calcutta	1,210		
Ships 'Rambler' and 'Millwood,' at sea, 1851; between lat. 16° 30′ N.L., 54° 30′ W., and lat. 23° 30′ N.L., 58° 0′ W.	1,056	Sea bottom resting on unknown rock	'Nautical Magazine'

THE LISBON EARTHQUAKE OF NOVEMBER 1, 1755.

Localities	Moment observed of shock		Distance from presumed origin		Velocity in feet per second
	h.	m.	°	′	
Presumed focus of shock, lat. 30°, long. 11° W.	9	23	—		—
A ship at sea in lat. 38°, long. 10° 47′ W.	9	24	0	30	3,091
Colares . .	9	30	1	30	1,325
Lisbon	9	32	1	30	1,030
Oporto	9	38	2	30	1,030
Ayamonte . . .	9	50	4	0	916
Cadiz	9	48	5	0	1,236
Tangier and Tetuan . .	9	46	5	30	1,478
Madrid	9	43	6	0	1,855
Funchal	10	1	8	30	1,382
Portsmouth . . .	10	3	12	30	1,431
Havre	10	23	13	0	1,339
Reading	10	27	13	30	1,304
Yarmouth	10	42	15	0	1,174
Amsterdam . . .	10	6	17	0	2,444
Loch Ness . . .	10	42	18	0	1,409

THE LISBON EARTHQUAKE OF MARCH 31, 1761.[1]

Locality	Moment observed of shock		Distance from presumed origin		Velocity in feet per second
	h.	m.	°		
Presumed focus, lat. 43°, long. 11° W.	11	51	—		—
Ship at sea in lat. 43°, many leagues from coast of Portugal	11	52	0	30	3,091
Ship in lat. 44° and about 80 leagues from coast	11	54	1	45	3,607
Corunna	11	51	2	30	2,576
Ship lat. 44° 8′ and 80 leagues NNW. of Cape Finisterre . . .	11	58	3	30	3,091
Lisbon	noon		4	30	3,091
Madeira	12	6	10	0	4,122
Cork	12	11	9	30	2,937

English miles each, and the time is reduced to Lisbon time.

These tables, owing to the nature of the materials

[1] See *Report of British Association*, 1858, p. 10.

which Mallet had at his disposal, are but rude approximations to the truth. Two interesting facts are, however, observable: the first being that the velocities for the earthquake of 1761 are much higher than those obtained for the earthquake of 1755; and, secondly, that in both cases the velocities as determined from the observations of ships at sea closely approximate to each other, in all cases being nearly the same as that with which a sound wave would travel through water.

The great differences in transit velocity obtained for different earthquakes is a point worthy of attention.

Seebach's velocity is a *true* transit velocity, and its determination is dependent on the assumption that the shock radiated from the *centrum* and not from the *epicentrum*. Seebach's method is explained when speaking about the determination of origins.

Some interesting observations on the velocity with which the earthquake of October 7, 1874, was propagated, are given by M. S. di Rossi.[1]

One assumption is that the disturbance radiated from an origin to surrounding points of observation, whilst another is that the disturbance followed natural fractures, the direction of which is derived from the crest of certain mountain ranges. These velocities are as follows, Maradi being at or near the origin of the disturbance :—

Velocity in feet per second with direct radiation			Velocity in feet per second by propagation along mountain chains			
Modigliana	.	. 820	By the Valley of	Marenzo	.	1,080
Bologna	.	. 656	,,	,,	Saveno	. 1,080
Forli .	.	. 874	,,	,,	Montone	. 1,080
Modena	.	. 518	,,	,,	Panaro	. { 1,080 / 984
Firenze	.	. 273	,,	,,	Sieve .	. 540
Compiobbi	.	. 328	,,	,,	,, .	. 540

[1] *Meteorologia Endogena*, i. p. 306.

Another set of interesting results are those of P. Serpieri on the earthquake of March 12, 1873. The curious manner in which this shock radiated is described in the chapter on the Geographical Distribution of Earthquakes (see p. 231). Two large areas appear to have been almost simultaneously struck, so that, there being no time for elastic yielding, the velocities calculated between places situated on either of the areas are exceedingly great.[1]

From Ragusa to Venice the velocity was 2,734 feet per second
" Spoleto " " 4,101 " "
" Perugia to Orvieto " 601 " "
" " " Ancona " 1,640 " "
" " " Rome " $\begin{cases} 1,640 \\ \text{or } 2,186 \end{cases}$ " "

The following are examples of approximate earthquake velocities which have been determined in Japan.

The Tokio Earthquake of October 25, 1881.—From records respecting this earthquake it appears to have been felt over the whole of Yezo and the northern and eastern coast of Nipon, a little farther south than Tokio. It was severest at Nemuro and Hakodate, and at the former place a little damage was done. From these facts, together with the indications of instruments recording direction of movement and a general inspection of the time records, it seems that the disturbance must have originated beneath the sea on the east coast of Yezo at a very long distance to the north-east of Tokio, from which place it passed in a practically direct line on to Yokohama.

As the disturbance was felt at Yokohama twenty-one seconds later than at Tokio, and the distances between these two places is about sixteen geographical miles, for this portion of its course the disturbance must have travelled at a rate of at least 4,300 *feet per second.* If

---

[1] See remarks on the Earthquake ' Push,' p. 162.

we assume that the shock, after having reached Hákodate, travelled on at the same rate as it did between Tokio and Yokohama in order to reach Saporo, where the shaking was felt eighteen seconds after Hakodate, it must have had about thirteen geographical miles to travel after Hakodate was shaken before Saporo felt its effect.

Drawing from Hakodate a tangent to the eastern side of a circle of thirteen miles radius described round Saporo, the origin of the disturbance must be on the line bisecting this tangent at right angles. As it also lies on a line drawn through Tokio and Yokohama, it lies in a position about 41 N. lat. and 144° 15′ E. long., which is a position somewhat nearer to Nemuro than Hakodate, as we should anticipate. If this be taken as approximately indicating the origin, then the shock, after reaching Hakodate from the Hakodate *homoseist*, travelled about 218 miles to reach Tokio in 128 seconds, which gives a *velocity of* 10,219 *feet per second*.

The method here followed is equivalent to that of the hyperbola and one direction (see p. 204). The hyperbola is described on the assumption that the velocity deduced from the time taken to travel between Tokio and Yokohama is correct, and also that the earth waves travelled with approximately the same velocity in the vicinity of Saporo as near Tokio. The probability, however, is that they travelled more quickly. If this be so, then the origin is thrown somewhat to the south-east and the velocity between the Hakodate homoseist and Tokio reduced. Thus, if the velocity in the Saporo district be double that observed in the Tokio district, the origin is shifted about twenty-eight miles to the south-west, and the last-mentioned velocity is reduced to about 9,000 feet per second.

If we work by the method of circles, and assume the velocity to have been constant in all directions, then this velocity must have been about 6,000 feet per second. If we assume that the indications of direction obtained from seismographs and other sources give to us by this intersection a proper origin, the velocity in some directions may have been as much as 17,000 feet per second.

An origin thus determined, or even if determined by the method of circles, is in discord with the fact that places like Nemuro, in the north-east of Yezo, were nearer to the origin than any of the other places which have been mentioned.

The conclusion which we are therefore led to with regard to this shock, assuming of course that the time observations are tolerably correct, is that the velocity of propagation was variable, being greater when measured between points near to the origin than between points at a distance. The velocities estimated vary between 4,000 and 9,000 feet per second.

In the case of the earthquake which has just been discussed, we have an example of a disturbance which must have passed between Tokio and Yokohama in what was almost a straight line from the origin. As this direction ought to give the maximum time of transit if all earthquakes are propagated with the same velocity, the following table is given of the interval between the time of observation of several shocks at these two stations :—

FROM YOKOHAMA TO TOKIO.			FROM TOKIO TO YOKOHAMA.		
1880 December 20th	36	seconds	1882 October 25th .	21	seconds
1881 January 7th	14–31	„	1883 February 6th .	23	„
„ March 8th .	60	„	„ March 1 .	53	„
„ „ 17th .	66	„	„ „ „ .	63	„
„ November 15th	31	„	„ „ 8th .	27	„
1882 February 16th	22	„	„ „ 11th .	26	„

As these are observations which have been made with the assistance of a telegraphic signal daily employed to correct and rate the clocks from which the observations were obtained, they may be regarded as being tolerably correct.

The disturbance of February 6, the two shocks of March 1, appear, like that of October 25, to have passed in almost a direct line from an origin in the N.N.E. through Tokio on to Yokohama. Their velocities of propagation as calculated from the above intervals are approximately 3,900, 1,900, and 1,400 feet per second. The shock of February 16 appears to have had its origin near to a point in Yedo Bay about eight miles east of Yokohama. Assuming this to be the case, the shock between the Yokohama homoseist and Tokio travelled at the rate of 2,454 feet per second, but between the Tokio homoseist and Chiba at the rate of 750 feet per second; that is to say, the velocity of propagation rapidly decreased as the disturbance spread outwards.

At Yokohama it was recorded at 5.31.54, at Tokio at 5.32.16, and at Chiba at 5.33.48. These times are given in Tokio mean time.

The shock of March 11, which was recorded at Tokio at 7.51.22 P.M. and at Yokohama at 7.51.33 P.M., appears, from the indications of instruments which were exceptionally definite in their records, to have originated in the N.E. corner of Yedo Bay, about nineteen miles S.S.W. from Chiba. This shock was rather severe, fracturing several chimneys. From the Tokio homoseist it appears to have travelled on to Yokohama at the rate of about 2,200 feet per second. Assuming these observations to be *approximately* accurate, if we take them with the records of previous observers they lead us to the following conclusions:—

1. Different earthquakes, although they may travel across the same country, have very variable velocities,

varying between several hundreds and several thousands of feet per second.

2. The same earthquake travels more quickly across districts near to its origin than it does across districts which are far removed.

3. The greater the intensity of the shock the greater is the velocity.

CHAPTER VI.

EFFECTS PRODUCED BY EARTHQUAKES UPON BUILDINGS.

The destruction of buildings is not irregular—Cracks in buildings—
Buildings in Tokio—Relation of destruction to earthquake motion—
Measurement of relative motion of parts of a building shaken by an
earthquake—Prevention of cracks—Direction of cracks—The pitch
of roofs—Relative position of openings in a wall—The last house
in a row—The swing of buildings—Principle of relative vibrational
periods.

THE subject of this chapter is, from a practical point of
view, one of the most important with which a seismologist
has to deal. We cannot prevent the occurrence of earthquakes, and unless we avoid earthquake-shaken regions,
we have not the means of escaping from them. What we
can do, however, is in some degree to protect ourselves.
By studying the effects produced by earthquakes upon
buildings of different construction and variously situated,
we are taught how to avoid or at least to mitigate calamities
which, in certain regions of the world, are continually
repeated. The subject is an extensive one, and what is
here said about it must be regarded only as a contribution
to the work of future writers who may give it the attention it deservedly requires.

*The Destruction produced by Earthquakes is not
irregular.*—If we were suddenly placed amongst the
ruins of a large city which had been shattered by an
earthquake, it is doubtful whether we should at once

recognise any law as to the relative position of the masses of *débris* and the general destruction with which we are surrounded. The results of observation have, however, shown us that, amongst the apparently chaotic ruin produced by earthquakes, there is in many cases more or less law governing the position of bodies which have fallen, the direction and position of cracks in walls, and the various other phenomena which result from such destructive disturbances.

Mallet, at the commencement of his first volume, describing the Neapolitan earthquake of 1857, discusses the general effect produced by various shocks upon differently constructed buildings. First he shows that, if we have a rectangular building, the walls at right angles to the shock will be more likely to be overthrown than those which are parallel to it. Experience teaches a similar lesson. Thus Darwin, when speaking of the earthquake at Concepcion in 1835,[1] tells us that the town was built in the usual Spanish fashion, with all the streets running at right angles to each other. One set ranged S.W. by W. and N.E. by E. and the other N.W. by N. and S.E. by S. The walls in the former direction certainly stood better than those in the latter. The undulations came from the S.W.

In Caraccas it is said that every house has its *laga securo*, or safe side, where the inhabitants place their fragile property. This *laga securo* is the north side, and it was chosen because about two out of every three destructive shocks traversed the city from west to east, so that the walls in these sides of a building have been stricken broadside on.[2]

---

[1] See *Researches in Geology and Natural History*, p. 374.
[2] 'The City of Earthquakes,' H. D. Warner, *Atlantic Monthly*, March, 1883.

Cracks in Buildings.—Results like the above come from destructive earthquakes rather than from movements such as those we have to deal with ordinarily. When a building is subjected to a slight movement it is assumed that the walls at right angles to the direction of the shock move backwards and forwards as a whole, and there is little or no tendency for them to be fractured at their weaker parts, these weaker parts being those over the various openings. The walls, however, which are parallel to the direction of the movement are, so to speak, extended and contracted along their length, and in consequence they may be expected to give way over the various openings. This tendency for extension and contraction of a wall along its length may be supposed, for instance, to be due to the different portions of a wall, owing to differences in dimensions and elasticity, having different periods of natural vibration, or possibly for two portions of a long line of wall to be simultaneously affected by portions of waves in different phases.

As an illustration of the giving way of a building in the manner here suggested we may take the case of a large brick structure which was recently being erected in Tokio. This building, at the time of the earthquake, was only some fourteen or fifteen feet above the surface of the ground. The length of the building stretched from N.W. to S.E., and it was intersected by many walls at right angles to this direction. Through all the walls of this building there were many arched openings. In the central part of the transverse walls, which walls were fully five feet in thickness, the arches which joined them together were 4 feet 4 inches in thickness. The arches therefore formed a comparatively lightly constructed link between heavy masses of brickwork.

On March 3, 1879, at 4.43 P.M., an earthquake was

felt throughout Tokio, the strength of which, as judged by our feelings, was above that of an average shock. As registered by one of Palmieri's instruments, it had a direction S.S.W. to N.N.E. and an intensity of 11°. On the same day there were several smaller shocks having the same direction, and these were succeeded by others on the 9th of the month.

Immediately after these shakings it was discovered that almost every arch in the internal walls of the building here referred to had been cracked across the crown in a direction about N. 40° W. All the other arches of the building, of which there were a great number in walls at right angles to the direction of the shock, were found not to have sustained any injury. To this statement, however, there was one exception, which was subsequently proved to have been due to a settlement taking place.

After examining these cracks the only cause to which they could be attributed was the series of shakings which they had just experienced. It seemed as if the heavy walls right and left of the arches had been in vibration without synchronism in their periods, and as a consequence the arches which connected them had been torn asunder.

Although the time at which the cracks were formed and the peculiar positions in which they were only to be found pointed distinctly to their origin, to be certain that they were not due to settlement of the foundations, horizontal lines were ruled upon the brickwork and from time to time subsequently observed.

The points to which the various cracks extended were also marked and observed. Beneath the walls as foundations there were beds of concrete about three feet thick and about ten feet in width. These had been under the pressure of the partially built walls for two years before the arches had been put in. As these foundations were unusually strong, being intended to carry so very much greater weight than that to which they had been subjected, if any settlement had been detected it would have been a matter of surprise.

Some weeks after the formation of these cracks it was observed that they gradually closed. This was probably due to the gradual falling inwards of the two broken portions of the arch, their position when open being one of instability.

Had this building been more complete at the time of the shock, and the heavy walls been tied together at higher points, although the archways would have been points of weakness, it is quite possible that fracture would not have taken place. This illustration shows us that when a building is shaken in a definite direction there will be

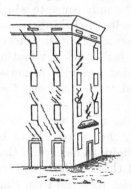

some rule as to the positions in which fractures occur. As another example, we may take the observations of Alexander Bittner upon the buildings of Belluno after the shock of June 29, 1873 (see Beiträge zur Kenntniss des Erdbebens von Belluno am 29 Juni, 1873, p. 40. Von Alexander Bittner. Aus dem LXIX. Bande der Sitzungsb. der K. Akad. der Wissensch., II. Abth., April-Heft. gahrg. 1874).

FIG. 16.—Cracks in a corner house, Belluno, June 29, 1873 (Bittner).

Speaking generally, he remarks that 'Houses similarly situated have suffered in corresponding walls and corners in a similar manner. In Belluno there is a certain kind of damage which is repeated everywhere, making a peculiar system of splits in the S.W. and N.E. corners of the houses.' This is well shown in the accompanying sketch, which evidently illustrates the effect of a shock oblique to the direction of two walls at right angles to each other.

Buildings in Tokio.—For the purpose of finding out what has been the effect produced by earthquakes upon the buildings of Tokio, and at the same time for ascertaining whether blocks of buildings ranging in different directions suffered to the same extent, the author examined, in company with Mr. Josiah Conder, a large number of foreign-built houses in the district of the Ginza. The chief reason for choosing this was because it was the only district where a large number of *similar* buildings could be found. By examining houses or buildings of different constructions, the effects produced upon them by earthquakes are very often likely to show so many differences that it becomes almost an impossibility to deter-

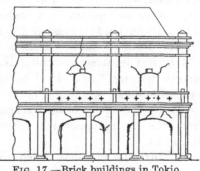

FIG. 17.—Brick buildings in Tokio, showing fractures.

mine what the general effect has been—unsymmetrical construction involving unsymmetrical ruin.

A number of similarly constructed buildings in one locality may be regarded as a number of seismographs, the effect upon any one of them being judged of by the average of the general effect which has been produced upon the whole. The general form of two of these houses which were examined is shown in fig. 17. In this figure the general character of the fractures which have been produced can also be seen. The houses are built of brick, and are in many cases faced with a thin coat of white plaster. Projecting from the level of the upper floor there is a balcony fronted by a low balustrade. This is supported by small beams which at their outer extremity are carried on a row of cylindrical columns. This forms a covered way in front of each row of houses. The roofs are covered with thick tiles. It will be observed that the arches of the upper windows spring *sharply* from their abutments, and at their crown they carry a heavy keystone. The lower openings, which have a span of 9 feet, have evidently been constructed in imitation of the open front of an ordinary Japanese house. These archways curve out *gently* from their abutments. The outside walls have a thickness of $13\frac{1}{2}$ inches.

The results obtained from a careful examination of 174 houses in streets running N.E. and 156 houses in streets running N.W., all of these houses being similar, were as follows :—

1. In the upper windows nearly all the cracks ran from the springing, which formed an angle with the abutment.

2. In the lower arches, which *curved* into the abutments, not a single crack was observed at the springway. The cracks in these arches were near the crown, where beams projected to carry the balcony. In many instances the cracks proceeded from such beams, even if there were no arch beneath. That cracks should occur in peculiar positions, as is here indicated, is shown in the illustrations which accompany the accounts of many earthquakes.

3. The houses which were most cracked were in the streets running parallel to the direction in which the greater number and most powerful set of shocks cross the city.

The results showed that, in order to avoid the effects of small shocks, all walls containing principal openings should be placed as nearly as possible at right angles to the direction in which the shocks of the districts usually travel. The blank walls, or those containing unimportant openings, would then be parallel to the direction of the shocks—that is, presuming our building to be made up of two sets of walls at right angles to each other.

Another point of importance would be to build archways *curving* into the supporting buttresses ; the archways over doors and windows which we find in earthquake countries do not appear to be in any way different from those which are built in countries free from earthquakes. In the one country these structures have simply to withstand vertical pressures applied statically ; in the other, they have to withstand more or less horizontal stresses, applied suddenly.

Relation of Destruction to Earthquake Motion.—The relations which exist between the overturning and projection of bodies and the motion of the ground have already been discussed. It may be interesting to call attention to the fact that in the formulæ showing three relationships, it was the *shape* rather than the *weight* of a body which determined whether it should be overturned or projected by a motion at its base.

As an interesting proof that light bodies may be overturned as easily as heavy ones, Mallet refers to the overturning of several large haystacks as one of the results of the Neapolitan earthquake.

If masses of material are displaced or fractured, then Mallet remarks that the maximum velocity will exceed $\sqrt{2gh}$, where h is the amplitude of the wave. Should the maximum velocity be less than this quantity, the masses which are acted upon will be simply raised and lowered, and there will be no relative displacements even if the emergence of the wave be nearly or quite vertical.

When we get a vertical wave acting upon an irregular mass of masonry, the heavier portions of the masonry, by their inertia, tend to descend relatively to the remaining portions, and in this way vertical fissures will be produced. For this reason it would not be advisable to use heavy materials above archways, heavy roofs, or heavy floors. The vertical fissures, Mallet remarks, would have their widest opening at the base.

In considering cases of fracture produced by earthquake motion, it must be remembered that these are due to stresses applied *suddenly*, and that if the same amount of stress had been *slowly* applied to a building, fractures might not have occurred.

If a disturbance is horizontal, and has a direction parallel to the length of a wall, the wall is carried forward at its foundations. This motion is opposed by the inertia of the upper portion of the wall and the various loads it carries. The wall being elastic, distortion takes place, and cracks, which are widest at the top, will be formed. In a uniform wall the two most prominent fissures ought to be near the ends.

If the horizontal backward and forward movement has a direction oblique to the plane of the wall, the wall will be either overthrown, fractured, or have a triangular fragment thrown off towards the origin from the end last reached.

Should the wave emerge steeply, diagonal fissures at right angles to the direction of transit will be formed, or else triangular pieces will be projected.

The accompanying figures are reduced from Mallet's 'Account of the Neapolitan Earthquake of 1857.'

Taking *a b* as the general direction of the fractures in

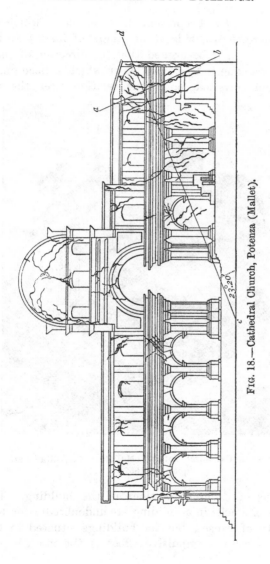

Fig. 18.—Cathedral Church, Potenza (Mallet).

fig. 18, then *c d* will represent the direction in which the
shock emerged, which is at an angle of 23°·20′ to the
horizon.   It might be argued that the direction of these
fractures was due to the direction in which surface undulations had travelled, or to the relative strengths and

FIG. 19.—The Cathedral, Paterno (Mallet).   Neapolitan Earthquake of 1857.

proportions of different portions of the building.   The
directions of cracks in a building are undoubtedly due to a
complexity of causes, but for buildings situated in the
region of shock the impulsive effect of the shock is pro-

bably the most important function to be considered. The method of applying the directions of emergence, deduced from observations on fractures, to determine the origin of a disturbance will be referred to in Chapter X.

Mallet observed that, although two ends of a building might be nearly the same, the fissures and joints do not occur at equal distances from the ends, nor are they equally opened.

The end where the joints are the most opened is that which was first acted upon, and this phenomenon may be sufficiently well pronounced to indicate the direction in which we must look to find the origin of a disturbance. Amongst possible explanations for this disposition of fractures in a wall, Mallet suggests that they may be due to real differences in the two semiphases of the wave of shock, the second semiphase being described with a somewhat slower velocity than the first. This, it will be observed, is contrary to the indications of seismographs.

Fig. 19, of the cathedral at Paterno, shows the effect of a subnormal shock striking a wall obliquely and projecting one of its corners.

MEASUREMENTS OF THE RELATIVE MOTION OF PARTS OF A
BUILDING AT THE TIME OF AN EARTHQUAKE.

In 1880 a series of observations was made in Tokio to determine whether at the time of an earthquake the various parts of the arched openings which we see in many buildings synchronised in their vibrations, or, for want of synchronism, were caused to approach and recede from each other. The arches experimented on were heavy brick arches forming the two corridors of the Imperial College of Engineering. The direction of one set of these corridors is N. 40° E. and that of the other N. 50° W.

The thickness of the walls in which these arches are placed is 1 ft. 11 in. They are built of Japanese bricks bound together with ordinary lime. The span of the arches is 8 ft. 3 in., and the height of the arch from the springing-line to the crown 4 ft. 1 in. The height of the abutments is 7 ft. 1½ in. The voussoirs of the arch are formed of a light grey soft volcanic rock, and on their faces show a depth of 12 inches. The width of the intermediate columns between the arches is 4 ft. 6⅞ in.

To determine whether at the time of an earthquake there was any variation in the dimensions of these arches, a light stiff deal rod, about 2 in. by ½ in. in cross section, was placed across the springing-line of the arch. One end of this was firmly fixed to the top of one abutment by means of a spike; on the other end, which was to indicate any horizontal movement if the abutments approached each other, there was fixed a pointer made out of a piece of steel wire. This rested on a piece of smoked glass fixed to the ledge on which the loose end of the rod was resting. If the abutments approached or receded from each other a line would be drawn measuring the extent of the motion. As a further indication of motion, a second smoked glass plate was fixed on the transverse rod, which plate was marked on by a pointer attached to a vertical rod hanging down from the crown of the arch.

As a general result of these experiments it may be said that the portions of the building which were examined usually either did not move at all, or else they practically synchronised in their movements. When they did move, the extent of motion was small, and the small differences in movement which were observed were in every probability far within the elastic limits of the structure.

Observations on Cracks.—To determine whether the walls of a building which have once been cracked, when subjected to a series of shocks, similar to those which they experienced before being cracked, still continued to give way, the extremities of a considerable number of cracks in the N.E. end of the museum buildings of the Engineering College were marked with pencil. Although since the time of marking there had been many severe shocks, these cracks did not visibly extend. These marks were made on the outside wall of the building. On the inside, one of these same cracks showed itself as a fissure about $\frac{1}{4}$ inch in width. Across this crack a horizontal steel wire pointer was placed. One end of this wire was fixed in the wall; the other end, which was pointed, rested on the surface of a smoked glass plate placed on the other side of the crack. After small earthquakes there was no indication of motion having taken place, but after a shock on February 21, as indicated by a line upon the smoked glass plate, it was seen that the sides of the crack had approached and receded from each other through a distance of about $\frac{1}{16}$ inch.

By similar contrivances placed on cracks in a neighbouring building exactly similar results were obtained, namely, that during small earthquakes the two sides of the crack had retained their relative positions, but at the time of a large shock this position had been changed.

In this building it was also observed that the cracks in many instances increased their length.

By attaching levers to the end of the pointers to multiply any motion that might take place, no doubt the indications would be more frequent and more definite. It would also be easier to note the relative distances of motion in two directions, namely, how far the cracks had closed and how far they had opened. As to whether motion would occur or not, much would no doubt depend upon the direction of the earthquake.

Prevention of Fractures.—One conclusion which may perhaps be drawn from these observations is, that a cracked building at the time of an earthquake shows a certain amount of flexibility. Whether a building which had been designed with cracks or joints between those parts which were likely to have different periods of vibration would be more stable, so far as earthquake shakings are concerned, than a similar building put up in an ordinary manner, is a matter to be decided by experiment. Certainly some of the cracks which have been examined indicate that if they had not existed, the strain upon the portion of the building where they occur would have been extremely great.

Direction of Cracks.—In looking at the cracks produced by small earthquakes it is interesting to note the manner of their extension. The basements of the buildings which have been most carefully examined are, for a height of two or three feet, built of large rectangular blocks of a greyish-coloured volcanic rock. In these parts the cracks pass in and out between the joints of the stone, indicating that the stones have evidently been stronger than the mortar which bound them together, and as a consequence the latter had to give way. Above this basement when the cracks enter the brickwork, they no longer exclusively confine themselves to the joints, but run in an irregular line through all they meet with, sometimes across the bricks and sometimes through the mortar joints. In places where they have traversed the brickwork, we can say that the mortar has been stronger than the bricks. This traversing of the bricks rather than the joints is, I think, the general rule for the direction of the cracks in the brickwork of Tokio buildings.

The Pitch of Roofs.—From observation of the effects

produced by earthquakes, it appears to us that the houses which lost the greater number of tiles appear to be those with the steepest pitch, and those where the tiles were simply laid upon the roof and not in any manner fastened down. It would seem that destruction of this sort might to a great extent be obviated by giving the roofs a less inclination and fixing the tiles with nails. It was also noticed that the greatest disturbance amongst the tiles was upon the ridges of the roofs. Destruction of this sort might be overcome by giving especial attention to these portions during the construction of the roof.

Relative Position of Openings in Walls.—From what has been said about the fractures in the buildings of Tokio it will have been seen that, with but few exceptions, they have all taken place above openings like doorways and windows. If architecture demands that openings like arches should be placed one above another in heavy walls of this kind, as in fig. 17, there will be lines of weakness running through the openings parallel to the dotted lines. As arches are only intended to resist vertical thrusts, special construction must be adopted to make them strong enough to resist horizontal pulls. For instance, a flat arch would offer more resistance to horizontal pulls than an arch put together with ordinary voussoirs, there being in the former case more friction to prevent the component parts sliding over each other. Or again, above each arch an iron girder or wooden lintel might be inserted in the brick or stone arch. It was suggested to me by my colleague, Mr. Perry, that the best form calculated to give a wall uniform strength, would be to build it so that the openings of each tier would occupy alternate positions, that is to say, along lines parallel to the struts and ties of a girder. In this way we should have our materials so arranged that they would offer the same resistance to hori-

zontal as to vertical movements.   Such a wall is shown in
fig. 20 : the dotted lines running through the openings, and
all similar lines parallel to the former, representing lines
of weakness.   If we
compare this with fig.
21, we shall see that in
the case of a horizontal
movement *a b* or of a
vertical movement *c d*,
we should rather ex-
pect to find fractures
in a house built like
fig. 21 than in one
built like fig. 20.   If,
however, these two buildings were shaken by a shock
which had an angle of emergence of about 45° in the
direction *e f*, the effects might be reversed.   Usually,
however, and always in a town like Tokio which is visited
by shocks originating at a distance, the movements are
practically horizontal ones, and, therefore, buildings
erected on the principles illustrated by fig. 20 should be
much superior, so far as resisting earthquakes is concerned,
to buildings constructed in the ordinary manner, as in
fig. 21.   Fractures following a vertical line of weakness are
shown in the accompanying drawing, fig. 22, of the Church
of St. Augustin, at Manilla, shattered by the earthquakes
of 1880.

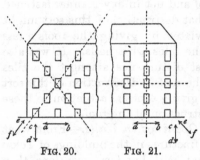

FIG. 20.   FIG. 21.

The last House in a Row.—When an earthquake shock
enters a line of buildings, and proceeds in a direction
coincident with that of the buildings, we should expect
that the last of these houses, being unsupported on one
side, would be in the position of the last person in Tyndall's
row of boys.   From this it would seem that the end house
in a row would show the greatest tendency to fly away
from its neighbours.   If the last house stood upon the

edge of a deep canal or a cliff, there would be a layer of
ground, equal in thickness to the depth of the canal or to
the height of the cliff, as the case may be, which would
also be in a position to be thrown forward.   The effect

FIG. 22.—Church of St. Augustin, Manilla.   Earthquakes of
July 18–20, 1880.

which is sometimes produced upon an end building is
shown in fig. 23, which is taken from the photograph of a
house shattered in 1868 at San Francisco.

I

The Swing of Buildings.—The distance through.
which buildings are moved at the time of an earthquake
depends partly on their construction and partly on the
extent, nature, and duration of the ·movement communicated to them at their foundations.   By violent
shocks buildings may be completely overthrown.   In the
case of small earthquakes, the upper portion of a house
may frequently move through a much greater distance
than the ground at its foundation.   For instance, during

FIG. 23.—Webber House, San Francisco.   Oct. 21, 1868.

the Yokohama earthquake of February, 1880, when the
maximum amplitude of the earth's motion was probably
under ¾ of an inch, from the slow swing of long Japanese
pictures, from three to six feet in length, which oscillated
backwards and forwards on the wall, it is very probable
that the extent through which the upper portion of
houses moved was very considerable.   In some instances
these pictures seem to have swung as much as two feet,
and from the manner in which they swung they evidently synchronised with the natural swing of the house.

From this it would seem that such a house must have rocked from side to side one foot out of its normal perpendicular position. That the motion was great is testified by nearly all who tried to stand at the time of the shock, it having been impossible to walk steadily across the floor of a room in an upper story. The houses here referred to are either those which are purely Japanese, or else those which are framed of wood and built on European models, a class of building which is very common in Tokio and Yokohama.

Perry and Ayrton calculated the period of a complete natural vibration of different structures. For a square house whose outer and inner sections were respectively 30 and 26 feet, and whose height was 30 feet, the period calculated would be about ·06 second.

At the time of the above earthquake many houses seem to have moved like inverted pendulums. On the morning after the shock my neighbour, who was living upstairs in a tall wooden house with a tile roof, told me that he endeavoured to count the vibrations, and was of the impression that to make a complete swing it took about 2 seconds.

Assuming now that the distance through which the top of a wooden house moved was about 1 foot, and the number of vibrations which it made per second was about ·5, then the greatest velocity of a point on the top of such a house must have been about 6 feet per second.

Mallet, who made observations upon the vibrations of various structures, tells us that Salisbury spire moves to and fro in a gale more than 3 inches. A well-constructed brick and mortar wall, 40 feet high and 1 foot 6 inches thick, was observed to vibrate in a gale 2 feet transversely before it fell.

An octagonal chimney with a heavy granite capping,

160 feet high, was observed instrumentally to vibrate at the top nearly 5 inches.[1]

At the time of a severe earthquake it does not seem impossible but that a building may be swung completely over. The accompanying illustration, fig. 24, taken from a photograph,[2] apparently indicates a movement of this description.

FIG. 24.—Stud Mill at Haywards, California.   Oct. 21, 1868.

Principle of relative Vibrational Period.—If a lath or thin pole loaded at one end with a weight fixed to the ground, so as to stand vertically, be shaken by an earthquake it will be caused to rock to and fro like an inverted pendulum. The period of its swing will be chiefly dependent on its dimensions, its elasticity, and its load. In a building we have to consider the vibration of a number of parts, the periods of which, if they were

---

[1] Mallet, *Dynamics of Earthquakes.*   [2] Stud Mill at Haywards.

independent of each other, would be different. On
account of this difference in period, whilst one portion of
a building is endeavouring to move towards the right,
another is pulling towards the left, and, in consequence,
either the bonds which join them or else they themselves
are strained or broken. This was strikingly illustrated
by many of the chimneys in the houses at Yokohama,
which by the earthquake of February 20, 1880, were
shorn off just above the roof. The chimneys were shafts of
brick, and probably had a slower period of vibration than
the roof through which they passed, this latter vibrating
with the main portion of the house, which was framed of
wood.

A particularly instructive example of this kind which
came under my notice is roughly sketched in fig. 25.

This is a chimney standing alone, which, for the sake
of support, was strapped by an iron band to an adjoining
building. It would seem that at
the time of the shock, the building
moving one way and the chimney
another, the swing of the heavy
building gave the chimney a sharp
jerk and cut it off. The upper
portion, being then loose upon the
lower part, rotated under the influence of the oscillations in manner
similar to that in which gravestones are rotated.

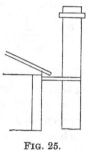

FIG. 25.

Mallet made observations similar to these in Italy.
He tells us that a buttress may often not have time to
transmit its stability to a wall. The wall and the
buttress have different periods of vibration, and therefore
they exert impulsive actions on each other. Effects like
these were strikingly observable in many of the rural

Italian churches where the belfry tower is built into one of the quoins of the main rectangular building.

Not only have we to consider the relative vibrations of the various parts of a building amongst themselves, but we have to consider the relation of the natural vibrations of any one of them or the vibration of the building as a whole, with regard to the earth, the vibrations of which it must be remembered are not strictly periodic.

Some of the more important results dependent upon the principle of 'relative vibrational periods' may be understood from the following experiments :—

In fig. 26 A, B, and C are three flat springs made out of strips of bamboo, and loaded at the top with pieces of

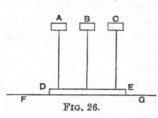

FIG. 26.

lead. At the bottom they are fixed into a piece of board D E, and the whole rests on a table F G. The legs of this table being slightly loose, by placing the fingers on the top of it, a quick short backward and forward movement can be produced. The weights on A and B are the same, but they are larger than the weight on C. Consequently the periods of A and B are the same, but different to the period of C. The dimensions of these springs are as follows : height, 18 inches; A and B each carry weights equal to 320 grammes, and they make one vibration per second; C has a weight of 199 grammes, and makes 0·75 vibrations per second.

First Experiment.—It will be found that by giving the table a gentle backward and forward movement, the extent of which movement may be so small that it will be difficult to detect it with the eye, either A and B may be made to oscillate violently whilst c remains still; or *vice versâ*, c may be caused to oscillate whilst A and B remain still. In the one case the period of shaking will have been synchronous with the natural period of A and B, whilst in the latter it will have been synchronous with that of c. This would seem to show us that if the natural period of vibration of a house, or of parts of it, at any time agree with the period of the shock, it may be readily thrown into a state of oscillation which will be dangerous for its safety.

Second Experiment.—Bind A and B together with a strip of paper pasted between them. (The paper used was three-eighths of an inch broad and would carry a weight of nearly three pounds.) If the table be now shaken as before, A and B will always have similar movements, and tend to remain at the same distance apart, and as a consequence the strip of paper will not be broken. From this experiment it would seem that so long as the different portions of a building have almost the same periods of vibration, there will be little or no strain upon the tie-rods or whatever contrivance may be used in connecting the different parts.

Third Experiment.—Join A and C, or B and C with a strip of paper in a manner similar to the last experiment. If the table be now shaken with a period approximating either to that of A and B, or with that of c, the paper will be suddenly snapped.

This indicates that if we have different portions of a building of such heights and thicknesses that their natural periods of vibration are different, the strain upon the portions which connect such parts is enormous, and it would seem, as a consequence, that either the vibrators themselves, or else their connections, must, of a necessity, give way. This was very forcibly illustrated in the Yokohama earthquake of February 1880 by the knocking over of chimneys. The particular case of the chimneys is, however, better illustrated by the next experiment.

Fourth Experiment.—Take a little block of wood three-quarters of an inch square and about one inch high, and place it on the top of A, B, or C. It will be found that, although the spring on which it stands is caused to swing backwards and forwards through a distance of three inches, the little block will retain its position.

This little block we may regard as the upper part of a chimney standing on a vibrating stack, and we see that, so long as this upper portion is light, it has no tendency to fall.

Fifth Experiment.—Repeat the fourth experiment, having first placed a small leaden cap on the top of the block representing the chimney. (The cap used only weighed a few grammes.) When vibration commences it will be found that the block quickly falls. This would seem to indicate that chimneys with heavy tops are more likely to fall than light ones.

Sixth Experiment.—Bind A and B together with a strip of paper and stand the little block upon the top of either. It will be found that the block will stand as in the fourth experiment.

Seventh Experiment.—Bind A and C, or B and C together, and place the block upon the top of either of them. When vibration commences, although the paper may not be broken, the little block will quickly fall.

Eighth Experiment.—Take two pencils or pieces of glass tube and place them under the board D E. If the table F G be now shaken in the direction D E, it will be found that the springs will not vibrate.

In a similar manner if a house or portion of a house were carried on balls or rollers, as has already been suggested, it would seem that the house might be saved from much vibration.

Ninth Experiment.—Set any of the springs in violent vibration by gently shaking D E instead of the table, and then suddenly cease the actuating motion. It will be observed that at the moment of cessation the board and the springs will have a sudden and very decided motion of translation in the same direction as that in which the springs were last moving, and although the springs were at the time swinging through a considerable arc, all motion will suddenly cease.

This shows, that if a house is in a state of vibration the strain at the foundations must be very great.

It would not be difficult to devise other experiments to illustrate other phenomena connected with the principle of relative vibrational periods, but these may perhaps be sufficient to show to those who have not considered this matter its great importance in the construction of buildings. Perhaps the greater portion of what is here said may by many be regarded as self-evident truisms hardly worth the trouble of demonstration. Their importance, however, seems to be so great that I hope that their discussion has not been altogether out of place.

I may remark that in the rebuilding of chimneys in Yokohama the principles here enunciated were taken advantage of by allowing the chimneys to pass freely through the roofs without coming in contact with any of the main timbers.

In putting up buildings to resist the effects of an earthquake, besides the idea of making everything strong because the earthquake is strong, there are several principles which, like the one just enunciated, might advantageously be followed which as yet appear to have received but little attention.

CHAPTER VII.

EFFECTS PRODUCED UPON BUILDINGS (*continued*).

Types of buildings used in earthquake countries—In Japan, in Italy,
in South America, in Caraccas—Typical houses for earthquake
countries—Destruction due to the nature of underlying rocks—
The swing of mountains—Want of support on the face of hills—
Earthquake shadows—Destruction due to the interference of waves
—Earthquake bridges—Examples of earthquake effects—Protection
of buildings—General conclusions.

Types of buildings used in earthquake countries.—
In Japan there are excellent opportunities of studying
various types of buildings. The Japanese types, of course,
form the majority of the buildings. The ordinary Japanese
house consists of a light framework of 4 or 5 inch scan-
tling, built together without struts or ties, all the timbers
crossing each other at right angles. The spaces are filled
in with wattle-work of bamboo, and this is plastered over
with mud. This construction stands on the top of a row
of boulders or of square stones, driven into the surface soil
to a distance varying from a few inches to a foot. The
whole arrangement is so light that it is not an uncommon
thing to see a large house rolled along from one position to another on wooden rollers. In buildings such as
these after a series of small earthquake shocks, we could
hardly expect to find more fractures than in a wicker
basket.

The larger buildings, such as temples and pagodas, are

also built of timber.   These are built up of such a multitude of pieces and framed together in such an intricate manner that they also are capable of yielding in all directions.   The European buildings are, of course, made of brick and stone with mortar joints.   Some of these, as the buildings of the Ginza in Tokio, are not designed for great strength.   On the other hand, others have thick and massive walls and are equal in strength to those we find in Europe.

The third type of buildings are those which are built in blocks ; and these blocks being bound together with iron rods traversing the walls in various directions are especially designed to withstand earthquakes.   A system somewhat similar to this has been patented in America, and examples of these so-called earthquake-proof buildings are to be found in San Francisco.

Speaking of Japanese buildings, Mr. R. H. Brunton, who has devoted especial attention to them says that,[1] 'to imagine that slight buildings, such as are seen here (i.e. in Japan), are the best calculated to withstand an earthquake shock is an error of the most palpable kind.'   After describing the construction of a Japanese house in pretty much the same terms as we have used, he says 'that with its unnecessarily heavy roof and weak framework it is a structure of all others the worst adapted to withstand a heavy shock.'   He tells us, further, that these views are sustained by the truest principles of mechanics.   In order to render buildings to some extent proof against earthquakes, some of the heavy roofs in Tokio have been so constructed that they are capable of sliding on the walls. Mr. Brunton mentions a design for a house, the upper

[1] See 'Constructive Art in Japan,' by R. H. Brunton, C.E., F.R.G.S., F.G.S., *Transactions of Asiatic Society of Japan*, December 22, 1873, and January 13, 1875.

part of which is to rest on balls, which roll on inverted
cups fixed on the lower part of the building, which is to be
firmly embedded in the earth. A similar design was,
at the suggestion of Mallet, used to support the tables
carrying the apparatus of some of the lighthouses erected
in Japan by Mr. Brunton. The very existence of these
designs seems to indicate that the ordinary European
house, however solidly and strongly it may be built, is
not sufficient to meet the conditions imposed upon it.
What is required, is something that will give way—an
approximation to the timber frame of a Japanese house,
so strongly condemned by Mr. Brunton and others. The
crucial test of the value of the Japanese structure, as compared with the modern buildings of brick and stone, is
undoubtedly to be found by an appeal to the buildings
themselves. So far as my own experience has gone, I must
say that I have never seen any signs in the Japanese
timber buildings which could be attributed to the effects
of earthquakes, and His Excellency Yamao Yozo, Vice
Minister of Public Works, who has made the study of the
buildings of Japan a speciality, told me that none of the
temples and palaces, although many of them are several
centuries old, and although they have been shaken by
small earthquakes and also by many severe ones, show
any signs of having suffered. The greatest damage wrought
by large earthquakes appears to have resulted from the
influx of large waves or from fires. In every case where
an earthquake has been accompanied by great destruction, by consulting the books describing the same,
it can be seen, from the illustrations in these books
portraying conflagrations, that this destruction was chiefly
due to fire. When we remember that nearly all Japanese
houses are constructed of materials that are readily
inflammable, it is not hard to imagine how destruction of

this kind has come about. To a Japanese, living as he does in a house which has been compared to a tinder-box, fire is one of his greatest enemies, and in a city like Tokio it is not at all uncommon to see during the winter months many fires which sweep away from 100 to 500 houses. In one winter I was a spectator of three fires, each of which was said to have destroyed upwards of 10,000 houses.

Although it would appear that the smaller earthquakes of Japan produce no visible effect upon the native buildings, it is nevertheless probable that small effects may have been produced, the observation of which is rendered difficult by the nature of the structure. If we look at buildings of foreign construction, by which are meant buildings of brick and stone, the picture before us is quite different, and everywhere the effects of earthquakes are palpable even to the most casual observer. Of these effects numerous examples have already been given. Not only are these buildings damaged by the cracking of walls and the overturning of chimneys, but they also appear to be affected internally. For instance, in the timbers of the roof of the museum attached to the Imperial College of Engineering in Tokio, there are a number of diagonal pieces acting as struts or ties intended to prevent more or less horizontal movements taking place. Those which are rigidly joined together with bolts and angle irons have apparently suffered from their rigidity, being twisted and bent into various forms. The buildings in Tokio, which are strongly put together, being especially designed to withstand earthquakes, appear to have suffered but little. I know only one example which at the time of the severe shock of 1880 had several of its chimneys damaged.

The ordinary houses in Italy, though built of stone and

mortar, are but poorly put together, and, as Mallet has remarked, are in no way adapted to withstand the frightful shakings to which they are subjected from time to time.

In the large towns, like Naples, Rome, and Florence, where happily earthquakes are of rare occurrence, although the building may be better than that found in the country, the height of the houses and the narrowness of the streets are sufficient to create a shudder, when we think of the possibility of the occurrence of a moderately severe earthquake.

In South America, although many buildings are built with brick and stone, the ordinary houses, and even the larger edifices, are specially built to withstand earthquakes. In Mr. James Douglas's account of a ' Journey Along the West Coast of South America,' we read the following[1] : ' The characteristic building material of Guayaquil is bamboo, which grows to many inches in thickness, and which, when cut partially through longitudinally at distances of an inch or so, and once quite through, can be opened out into fine elastic boards of serviceable width. Houses, and even churches, of a certain primitive beauty are built of such reeds, so bound together with cords that few nails enter into the construction, and which, therefore, yield so readily to the contortions of the earth during an earthquake as to be comparatively safe.'

Here we have a house, which, so far as earthquakes are concerned, is an exaggerated example of the principles which are followed in the construction of an ordinary Japanese dwelling.

Another plan adopted in South America can be gathered from the same author's writings upon Lima, about which he says, ' To build high houses would be to erect structures for the first earthquake to make sport of, and, therefore, in order to obtain space, safety, and comfort, the houses of the wealthy surround court after court, filled with flowers, and cooled with fountains, connected one with another with wide passages which give a vista from garden to garden.'

History would indicate that houses of this type have been arrived at as the results of experience, for it is said that when the inhabitants of South America first saw the Spaniards building tall houses, they told them they were building their own sepulchres.[1]

In Jamaica, we find that even as early as 1692 experience had taught the Spaniards to construct low houses, which withstood shakings better than the tall ones.[2]

In Caraccas, which has been called the city of earthquakes, it is said that the earthquakes cause an average yearly damage amounting to the equivalent of a *per capita* tax of four dollars. To reduce this impost to a minimum much attention is paid to construction. 'Projecting basement corners (giving the house a slightly pyramidal appearance) have been found better than absolutely perpendicular walls; mortised corner-stones and roof beams have saved many lives when the central walls have split from top to bottom; vaults and key-stone arches, no matter how massive, are more perilous than common wooden lintels, and there are not many isolated buildings in the city. In many streets broad iron girders, riveted to the wall, about a foot above the house door, run from house to house along the front of an entire square. Turret-like brick chimneys, with iron top ornaments, would expose the architect to the vengeance of an excited mob; the roofs are flat, or flat terraced; the chimney flues terminate near the eaves in a perforated lid.'[3]

[1] *Phil. Trans.*, li. 1760.   [2] *Ibid.*, xviii.
[3] 'The City of Earthquakes,' H. D. Warner, *Atlantic Monthly* March 1883.

Typical houses for earthquake countries.—From what has now been said about the different buildings found in earthquake countries, it will be seen that if we wish to put up a building able to withstand a severe shaking, we have before us structures of two types. One of these types may be compared with a steel box, which, even were it rolled down a high mountain, would suffer but little damage; and the other, with a wicker basket, which would equally withstand so severe a test. Both of these types may be, to some extent, protected by placing them upon a loose foundation, so that but little momentum enters them at their base. One suggestion is to place a building upon iron balls. Another method would be to place them upon two sets of rollers, one set resting upon the other set at right angles. The Japanese, we have seen, place their houses on round stones. The solid type of building is expensive, and can only be approached partially, whilst the latter is cheap, and can be approached closely. In the case of a solid building it would be a more difficult matter to support it upon a movable foundation than in the case of a light framework. Such a building is usually firmly fixed on the ground, and consequently at the time of an earthquake, as has already been shown by experiment, must be subjected to stresses which are very great. In consequence also of the greater weight of the solid structure, more momentum will enter it at its base than in the case of the light structure. Also, we must remember that the rigidity favours the transmission of momentum, and with rigid walls we are likely to have ornaments, coping-stones, and the comparatively freer portions forming the upper part of a building displaced; whilst, with flexible walls absorbing momentum in the friction of their various parts, such disturbances would not be so likely. Mr. T. Ronaldson, referring to

this, says, that in 1868, at San Francisco, the ornamental
stone-work in stone and cement buildings was thrown
from its position, whilst similar ornaments in neighbouring brick buildings stood.

To reduce the top weight of a building, hollow bricks
might be employed. To render a building more homogeneous and elastic, the thickness of bricks might be
reduced. Inasmuch as the elasticity of brick and timber
are so different, the two ought to be employed separately.
For internal decorations plaster mouldings might be re-
placed by *papier mâché* and *carton-pierre*, the elastic
yielding of which is comparatively great.[1] Houses,
whether of brick and stone, or of timber, ought to be
broad and low, and the streets three or four times as wide
as the houses. The flatter the roofs the better.

One of the safest houses for an earthquake country
would probably be a one-storied strongly framed timber
house, with a light flattish roof made of shingles or sheet-
iron, the whole resting on a quantity of small cast-iron balls
carried on flat plates bedded in the foundations. The
chimneys might be made of sheet-iron carried through
holes free of the roof. The ornamentation ought to be
of light materials.

At the time of severe earthquakes many persons seek
refuge from their houses by leaving them. In this case
accidents frequently happen from the falling of bricks
and tiles. Others rush to the doorways and stand beneath
the lintels. Persons with whom the author has conversed
have suggested that strongly constructed tables and bedsteads in their rooms would give protection. To see
persons darting beneath tables and bedsteads would undoubtedly give rise to humiliating and ludicrous exhibi-
tions. This latter idea is not without a value, and most

[1] T. Ronaldson, *A Treatise on Earthquake Dangers, &c.*

K

certainly, if applied in houses of the type described, would be valuable.

The great danger of fire may partially be obviated by the use of 'earthquake lamps,' which are so constructed that before they overturn they are extinguished. It is said that in South America some of the inhabitants are ready at any moment to seek refuge in the streets, and they have coats prepared, stocked with provisions and other necessaries, which, if occasion demands, will enable them to spend the night in the open air. These coats, called 'earthquake coats,' might also, with properly constructed houses, be rendered unnecessary.

Destruction due to the nature of the underlying rocks.—That the nature of the ground on which a building stands is intimately related with the severity of the blow it receives is a fact which has often been demonstrated.

One cause of destruction is due to placing a building on foundations which are capable of receiving the full effects of a shock, and transmitting it to the buildings standing on them.

For instance, the reason why a soft bed might possibly make a good foundation, is, as has been pointed out by Messrs. Perry and Ayrton, because the time of transmission of momentum is increased ; in fact, the soft bed is very like a piece of wood interposed between a nail and the blows of a hammer—it lengthens the duration of impact. For this reason we are told that a quaking bog will make a good foundation. When a shock enters loose materials its waves will be more crowded, and it is possible that a line of buildings may rest on more than one wave during a shock. There are many examples on record of the stability of buildings which rested on beds of particular material at the time of destructive earthquakes. As the observations which have been made by various writers on this subject appear to point in a contrary direction, I give the following examples :—

In the great Jamaica earthquake of 1692, the portions of Port Royal which remained standing were situated on a compact limestone foundation ; whilst those on sand and gravel were destroyed ('Geological Observer,' p. 426). Again, on p. 148 of the same work, we read, 'According to the observations made at Lisbon, in 1737, by Mr. Sharpe, the destroying effects of this earthquake were confined to the tertiary strata, and were most violent on the blue clay, on which the lower part of the city is constructed. Not a building on the secondary limestone or on the basalt was injured.'

In the great earthquakes of Messina, those portions of the town situated on alluvium, near the sea, were destroyed, whilst the high parts of the town, on granite, did not suffer so much. Similar observations were made in Calabria, when districts consisting of gravel, sand, and clay became, by the shaking, almost unrecognisable, whilst the surrounding hills of slate and granite were but little altered. At San Francisco, in 1868, the chief destruction was in the alluvium and made ground.

At Talacahuano, in 1835, the only houses which escaped were the buildings standing on rocky ground ; all those resting on sandy soil were destroyed.

From the results of observations like these, it would seem the harder rocks form better foundations than the softer ones. The explanation of this, in many cases, appears to lie in the fact that the soft strata were in a state of unstable equilibrium, and by shaking, they were caused to settle. Observations like the following, however, point out another reason why soft strata may sometimes afford a bad foundation.

'Humboldt observed that the Cordilleras, composed of gneiss and mica-slate, and the country immediately at their foot, were more shaken than the plains.'[1]

'Some writers have asserted that the wave-like movements (of the Calabrian earthquake in 1783) which were propagated through recent strata from west to east, became very violent when they reached the point of junction with the granite, as if a reaction was produced when the undulatory movement of the soft strata was suddenly arrested by the more solid rocks.'

Dolomieu when speaking of this earthquake says, the usual effect 'was to disconnect from the sides of the Apennines all those masses (of sand and clay) which either had not sufficient bases for their bulk, or which were supported only by lateral adherence.'

These intensified actions taking place at and near to lines of junction between dissimilar strata is probably due to the phenomena of reflection and refraction.

When referring to the question as to whether buildings situated on loose materials suffered more or less than those on solid rocks, Mallet, in his description of the Neapolitan earthquake of 1857, remarks: 'We have in this earthquake, towns such as Saponara and Viggiano, situated upon solid limestone, totally prostrated; and we have others such as Montemarro, to a great extent based upon loose clays, totally levelled. We have examples of almost complete immunity in places on plains of deep clay as that of Viscolione, and in places on solid limestone, like Castelluccio, or perched on mountain tops like Petina.'[2]

After reading the above, we see that the probable reason why, in several cases, beds of soft materials have not made

[1] *Principles of Geology*, Lyell, vol. ii. p. 106.
[2] *The Neapolitan Earthquake of* 1857, R. Mallet, vol. ii. p. 359.

good foundations, consists in the fact that they have either been of small extent or else have been observed only in the neighbourhood of lines which divided them from other formations, which lines are always those of great disturbances.

At the end of his description of the Neapolitan earthquake of 1857, Mallet says that more buildings were destroyed on the rock than on the loose clay. This, however, he remarks, is hardly a fact from which we can draw any valuable deductions, because it so happened that more buildings were constructed on the hills than on the loose ground.[1]

Professor D. S. Martin, writing on the earthquake of New England in 1874, remarks that in Long Island the shock was felt where there was gneiss between the drift. Around portions to the east the observations were few and far between. He also remarks that generally the shocks were felt more strongly and frequently on rocky than on soft ground.[1]

From these examples, it would appear that the hard ground, which usually means the hills, forms a better foundation than the softer ground, which is usually to be found in the valleys and plains. Other examples, however, point to a different conclusion. For instance, a civil engineer, writing about the New Zealand earthquake of 1855, when all the brick buildings in Wellington were overthrown, says that ' it was most violent on the sides of the hills at those places, and least so in the centre of the alluvial plains.' [2]

In this example it must be noticed that the soft alluvium here referred to was of large extent, and not loose material resting on the flanks of rocks, from which it was

[1] *Am. J. Sci.* x. 191.
[2] *Reports of British Association*, 1858, p. 106.

likely to be shaken down, as in most of the previous examples.

The results of my own observations on this subject point as much in one direction as in the other. In Tokio, from instrumental observations upon the slopes and tops of hills, the disturbance appears to be very much less than it is in the plains. Thus, at my house, situated on the slope of a hill about 100 feet in height, for the earthquake of March 11, 1882, I obtained a maximum amplitude of motion of from three to four millimètres only, whilst Professor Ewing, with a similar instrument, situated on the level ground at about a mile distant, found a motion of fully seven millimètres. This calculation has been confirmed by observations on other earthquakes. Thus, for instance, in the destructive earthquake of 1855, when a large portion of Tokio was devastated, it was a fact, remarked by many, that the disturbance was most severe on the low ground and in the valleys, whilst on the hills the shock had been comparatively weak. As another illustration, I may mention that within three-quarters of a mile from my house in Tokio there is a prince's residence which has so great a reputation for the severity of the shakings it receives that its marketable value has been considerably depreciated, and it is now untenanted.

In Hakodadi, which is a town situated very similarly to Gibraltar, partly built on the slope of a high rocky mountain and partly on a level plain, from which the mountain rises, the rule is similar to that for Tokio, namely, that the low, flat ground is shaken more severely than the high ground. At Yokohama, sixteen miles southwest from Tokio, the rule is reversed, as was very clearly demonstrated by the earthquake of February 1880, when almost every house upon the high ground lost its chimney, whilst on the low ground there was scarcely any damage done ; the only places on the low ground which suffered were those near to the base of the hills. The evidence as to the relative value of hard ground as compared with soft ground, for the foundation of a building, is very conflicting. Sometimes the hard ground has proved the better foundation and sometimes the softer, and the superiority of one over the other depends, no doubt, upon a variety of local circumstances.

These latter observations open up the inquiry as to the extent to which the intensity of an earthquake may be modified by the topography of the disturbed area.

The swing of mountains.—If an earthquake wave is passing through ground the surface of which is level, so long as this ground is homogeneous, as the wave travels further and further we should expect its energy to become less and less, until, finally, it would insensibly die out. If, however, we have standing upon this plain a mountain, judging from Mallet's remarks, this mountain would be set in a state of vibration much in the same way as a house is set in vibration, and it would tend to oscillate backward and forward with a period of vibration dependent upon the nature of its materials, size, and form. The upper portion of this mountain would, in consequence, swing through a greater arc than the lower portion, and buildings situated on the top of it would swing to and fro through a greater arc than thóse which were situated near its foot. This explanation why buildings situated on the top of a mountain should suffer more than those situated on a plain, is one which was offered by Mallet when writing of the Neapolitan earthquake. He tells us that towns on hills are ' rocked as on the top of masts,' and if we accept this explanation it would, in fact, be one reason why the houses situated on the Bluff at

Yokohama suffered more than those situated in the settlement. This explanation is given on account of the great
authority it claims as a consequence of its source. It is
not clear how the statement can be supported, as different
portions of the mountain receive momentum in opposite
directions at the same time.

Want of support on the faces of hills.—When a
wave of elastic compression is propagated through a
medium, we see that the energy of motion is being continually transmitted from particle to particle of that
medium. A particle, in moving forwards, meets with an
elastic resistance of the particles towards which it moves,
but, overcoming these resistances, it causes these latter
particles to move, and in turn to transmit the energy to
others further on. So long as the medium in which this
transfer of energy is continuous, each particle has a limit
to its extent of motion, dependent on the nature of the
medium. When, however, the medium, which we will
suppose to be the earth, is not continuous, but suddenly
terminates with a cliff or scarp, the particles adjacent to
this cliff or scarp, having no resistance offered to their
forward motion, are shot forward, and, consequently, the
ground here is subjected to more extensive vibrations
than at those places where it was continuous. This may
be illustrated by a row of marbles lying in a horizontal
groove ; a single marble rolled against one end of this
row will give a concussion which will run through the
chain, like the bumping of an engine against a row of
railway cars, and as a result, the marble at the opposite
end of the row, being without support, will fly off.
Tyndall illustrates the same thing with his well known
row of boys, each one standing with his arms stretched
out and his hands resting upon the shoulders of the boy
before him. A push being given to the boy at the back,

the effect is to transmit a push to the first boy, who, being unsupported, flies forward.

In the case of some earthquakes, most disastrous results have occurred which seem only to admit of an explanation such as this. A remarkable instance of this kind occurred when the great earthquake of 1857 'swept along the Alps from Geneva to the east-north-east, and its crest reached the edge of the deep glen between Zermatt and Visp. Then the upper part of the wave-movement, a thousand or two thousand feet in depth from the surface, came to an end ; the forward pulsation acted like the breaker of the sea, and heavy falls of rock encumbered the western side of the valley.'

Earthquake shadows.—If a mountain stands upon a plain through which an elastic wave is passing, which is almost horizontal, the mountain is, so to speak, in the *shadow* of such a wave. If we only consider the normal motion of this wave, we see that the only motion which the mountain can obtain will be a wave of elastic distortion produced by a shearing force along the plain of the base. Should, however, the wave approach the mountain from below, and emerge into it at a certain angle, only the portion of the mountain on the side from which the wave advanced could remain in shadow, whilst the portion on the opposite side would be thrown into a state of compression and extension. Portions in shadow, however, would be subject to waves of elastic distortion. In a manner similar to this we may imagine that certain portions of the bluff, so far as the advancing wave was concerned, were in shadow, and thus saved from the immediate influence of the direct shock. A hypothetical case of such a shadow is shown in the accompanying section, illustrating the contour of the ground at Yoko-hama. The situation which might be in the shadow of one shock, however, it is quite possible might 'not be in that of another. We must also remember that a place in

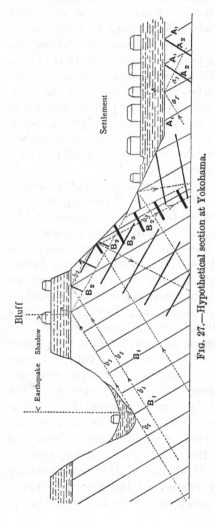

FIG. 27.—Hypothetical section at Yokohama.

shadow for a direct shock might be affected by reflected waves, and also by the transverse vibrations of the direct shock. These effects are over and above the effects produced by the waves of elastic distortion just referred to. It might be asked whether whole countries, like England, which are but seldom shaken, are in shadow.

Destruction due to the interference of waves.—Referring to the section of the ground at Yokohama (Fig. 27), it will be seen that both the settlement and 'the bluff stand upon beds of gravel capping horizontal beds of grey tuff. The gravel of that portion of the settlement on the seaboard originally formed the line of a shingle beach. That portion of the settlement back from the sea stands upon ground which was originally marshy. In the central portions of the settlement this bed of gravel is very thick, perhaps 100 feet or so, but as you near the edge of the bluff it probably becomes thinner, until it finally dies out upon the flanks of the scarps.

On the top of the bluff, the beds of gravel will, in every probability, be generally thinner than they are upon the lower level. The beds of tuff, which is a soft grey-coloured clay-like rock, produced by the solidification of volcanic mud, appear, when walking on the seaboard, to be horizontally stratified. If there is a dip inland, it is in all probability very slight. Here and there the beds are slightly faulted. Taken as a whole we may consider these beds as being tolerably homogeneous, and an earthquake in passing through them would meet with but little reflection or refraction. At the junction of these beds with the overlying gravels, both reflection and refraction would comparatively be very great.

On entering the gravel, as the wave would be passing into a less elastic medium, the direction of the wave would be bent towards the perpendicular to the line of junction, and the angle of emergence at the surface would consequently be augmented. At the surface certain reflection would also take place, but the chief reflections would be those at the junction of the tuff and the alluvium.

Under the settlement it is probable that all the reflections which took place would be single. Thus wave fronts like A_1 advancing in a direction parallel to the line a_1 would be reflected in a direction a_2 and give rise to a series of reflected waves A_2. These are shown by thicker lines. Similarly all the neighbouring waves to the right and left of A_1 would give rise to a series of reflected waves. If the lines drawn representing wave fronts are districts of compression, then, where two of the lines cross each other, there would be double energy in producing compression. Similarly, districts of rarefaction might accord, and, again, compression of one wave might meet with the rarefaction of another and a neutralisation of effect take place. A diagram illustrating concurrence and interference of this description is given in Le Conte's ' Elements of Geology,' p. 115. The interference which has been spoken of, however, is not the greatest which would occur. The greatest would probably be beneath the bluff and the scarps which run down to join the level ground below. This would be the case because it is a probability that there might not only be cases of interference of single reflected waves, but also of waves which had been not only twice but perhaps thrice reflected. For example, a wave like B_1 (which is parallel to A_1 of the first supposition), advancing in a direction parallel to b_1, might be reflected along the line b_2 giving rise to waves like B_2, which in turn might be reflected along b_3 giving rise to waves like B_3. The number of districts where there would be concurrence and interference would, in consequence of the number of times waves might be reflected, be augmented. Here the violence of the shock would, at certain points, be considerably increased, but as a general result energy must be lost, so that even if some of the reflected waves found their way into the portion we have regarded as being in shadow, their intensity would not be so great as if they had entered it directly.

The shaking down of loose materials from the sides of hills may be partially explained on the assumption of an increased disturbance due to interference.

Earthquake bridges.—In certain parts of South

America there appear to exist tracts of ground which are practically exempt from earthquake shocks, whilst the whole country around is sometimes violently shaken. It would seem as if the shock passes beneath such a district as water passes beneath a bridge, and for this reason these districts have been christened ' bridges.'

This phenomenon appears to depend upon the nature of the underlying soil. When an elastic wave passes from oné bed of rock to another of a different character, a certain portion of the wave is reflected, while the remainder of it is transmitted and refracted, and ' bridges ' we may conceive of as occurring where the phenomenon of total reflection occurs.

In the instances given of soft materials having proved good foundations, it was assumed that they had chiefly acted as absorbers of momentum. They have also acted as reflecting surfaces, and where no effects have been felt by those residing on them, this may have been the result of total reflection, and the soft beds thus have played the part of bridges.

Fuchs gives an example taken from the records of the Syrian earthquake of 1837, where not only neighbouring villages suffered differently, but even neighbouring houses. In one case a house was entirely destroyed, whilst in the next house nothing was felt.

In Japan, at a place called Choshi, about 55 miles east of the capital, earthquakes are but seldom felt, although the surrounding districts may be severely shaken.

From descriptions of this place it would appear that there is a large basaltic boss rising in the midst of alluvial strata. The immunity from earthquakes in this district has probably given rise to the myth of the Kanam rock, which is a stone supposed to rest upon the

head of a monstrous catfish (Namadzu), which by its writh-
ings causes the shakings so often felt in this part of the
world.[1]

Prof. D. S. Martin, writing on the earthquake of New
England in 1874, says that it was felt at four points; it
was felt in the heart of Brooklyn all within a circle of
half a mile across; 'and this fact would suggest that a
ridge of rock perhaps approaches the surface at that
point, though none is known to appear.'[2]

The subject of special districts, which are more or less
protected from severe shakings, will be again referred to,
and it will be seen that after a seismic survey has been
made even of a country like Japan, where there are on
the average at least two earthquakes per day, it is possible
to choose a place to build in as free from earthquakes as
Great Britain.

General examples of earthquake effects.—The following examples of earthquake effects are drawn from
Mallet's account of the Neapolitan earthquake of 1857.

At a town called Polla there was great destruction.
Judging from the fissures in the parts that remained
standing it seemed that the emergence of the shock had
been more vertical in the upper part of the town than in
the lower, proving that whatever had been the angle
below, the hill had itself vibrated, which, being horizontal,
had modified the angle of the fissures.

Diano suffered but little, partly because it was well
built, and partly on account of its situation, which was
such that before the shock reached it the disturbance
had to pass from beds of clay into nearly vertically placed
beds of limestone. Also a great portion of the shock was
cut off by the Vallone del Raccio to the north and north-

[1] See chapter 'Causes of Earthquakes' for details of this myth.
[2] *Am. Jou. Sci.* vol. x. p. 191.

west of the town. Here the effects of the partial extinction of the wave on the ' free outlaying stratum' were visible in the masses of projected rock.

Castellucio did not suffer because its well buttressed knoll was end on to the direction of shock, and on account of a barrier of vertical breccia beds protecting it upon the east.

Pertosa stands on a mound. The destruction was least in the southern part of the town. From the relation of the beds of breccia on which the town stands, and the direction of the wave path, it is evident that the southern part of the town received the force of the shock through a greater thickness of the breccia beds than the other parts did.

Petina, standing on a level limestone spur jutting out from a mountain slope, suffered nothing, whilst Anletta five miles to the south-west, and Pertosa six miles distant, were in great part prostrated. (1) The terrace did not vibrate, and (2) between Petina and Anletta there is almost 6,000 feet of piled up limestone, so that any shock emergent at a steep angle had to pass up transversely through these beds.

Protection of buildings.—In addition to giving proper construction to our buildings, choosing proper foundations and positions for them, something might possibly be done to ward off the destructive effects of an earthquake. We read that the Temple of Diana at Ephesus was built on the edge of a marsh, in order to ward off the effect of earthquakes. Pliny tells us that the Capitol of Rome was saved by the Catacombs, and Elisée Reclus [1] says that the Romans and Hellenes found out that caverns, wells, and quarries retarded the disturbance of the earth, and protected edifices in their neigh-

[1] *The Earth*, p. 599.

bourhood. The tower of Capua was saved by its numerous wells. Vivenzis asserts that in building the Capitol the Romans sunk wells to weaken the effects of terrestrial oscillations. Humboldt relates the same of the inhabitants of San Domingo.

Quito is said to receive protection from the numerous cañons in the neighbourhood, whilst Lactacunga, fifteen miles distant, has often been destroyed.

Similarly, it is extremely probable that many portions of Tokio have from time to time been protected more or less from the severe shocks of earthquakes by the numerous moats and deep canals which intersect it.

Although we are not prepared to say how far artificial openings of this description are effectual in warding off the shocks of earthquakes, from theoretical considerations, and from the fact that their use has been discovered by persons who, in all probability, were without the means of making theoretical deductions, the suggestions which they offer are worthy of attention.

General conclusions.—The following are a few of the more important results which may be drawn from the preceding chapter :—

1. In choosing a site for a house find out by the experience of others or experimental investigation the localities which are least disturbed. In some cases this will be upon the hills, in others in the valleys and on the plains.

2. A wide open plain is less likely to be disturbed than a position on a hill.

3. Avoid loose materials resting on harder strata.

4. If the shakings are definite in direction, place the blank walls parallel to such directions, and the walls with many openings in them at right angles to such directions.

5. Avoid the edges of scarps or bluffs, both above and below.

6. So arrange the openings in a wall, that for horizontal stresses the wall shall be of equal strength for all sections at right angles.

7. Place lintels over flat arches of brick or stone.

8. To withstand destructive shocks either rigidly follow one or other of the two systems of constructing an earthquake-proof building. The light building on loose foundations is the cheaper and probably the better.

9. Let all portions of a building have their natural periods of vibration nearly equal.

10. If it is a necessity that one portion of a building should have a very different period of vibration to the remainder, as for instance a brick chimney in a wooden house, it would seem advisable either to let these two portions be sufficiently free to have an independent motion, or else they must be bound together with great strength.

11. Avoid heavy topped roofs and chimneys. If the foundations were free the roof might be heavy.

12. In brick or stone work use good cement.

13. Let archways curve into their abutments.

14. Let roofs have a low pitch, and the tiles, especially those upon the ridges, be well secured.

L

CHAPTER VIII.

EFFECTS OF EARTHQUAKES ON LAND.

1. Cracks and fissures—Materials discharged from fissures—Explanation of fissure phenomena. 2. Disturbances in lakes, rivers, springs, wells, fumaroles, &c.—Explanation of these latter phenomena. 3. Permanent displacement of ground—On coast lines—Level tracts —Among mountains—Explanation of these movements.

Cracks and fissures formed in the ground.—Almost all large earthquakes have produced cracks in the ground. The cracks which were found in the ground at Yokohama (February 22, 1880) were about two or three inches wide, and from twenty to forty yards in length. They could be best seen as lines along a road running near the upper edge of some cliffs which overlook the sea at that place. The reason that cracks should have occurred in such a position rather than in others was probably owing to the greater motion at such a place, due to the face of the cliff being unsupported, and there being no resistance opposed to its forward motion. It often happens that earthquake cracks are many feet in width. At the Calabrian earthquake of 1783, one or two of the crevasses which were formed were more than 100 feet in width and 200 feet in depth. Their lengths varied from half a mile to a mile.[1] Besides these large cracks, many smaller ones of one or two feet in breadth and of great length were formed. In

---

[1] Lyell, *Principles of Geology*, vol. ii. chap. xxix.

the large fissures many houses were engulfed. Subsequent excavations showed that by the closing of the fissures these had been jammed together to form one compact mass. These cracks are usually more or less parallel, and at the same time parallel to some topographical feature, like a range of mountains. For example, the cracks which were formed by the Mississippi earthquake of 1812 ran from north-east to south-west parallel to the Alleghanies. By succeeding shocks these crevasses are sometimes closed and sometimes opened still wider. Their permanency will also depend upon the nature of the materials in which they are made.

During an earthquake large cracks may suddenly open and shut.

During the convulsions of 1692 which destroyed Port Royal, it is said that many of the fissures which were formed, opened and shut. In some of these, people were entirely swallowed up and buried. In others they were trapped by the middle, and even by the neck, where if not killed instantaneously they perished slowly. Subsequently their projecting parts formed food for dogs.[1]

The earthquake which, July 18, 1880, shook the Philippines caused many fissures to be found, which in some places were so numerous that the ground was broken up into steps. Near to the village of San Antonio the soil was so disturbed that the surface of a field of sugar-canes was so altered that in some cases the top of one row of full grown plants was on a level with the roots of the next. Into one such fissure a boat disappeared, and into another, a child.

Subsequently the child was excavated, and its body, which was found a short distance below the surface, was completely crushed.[2]

---

[1] *Gent. Mag.* vol. xx. p. 212.   [2] *Trans. Seis. Soc.* vol. v. p. 67–68.

At the time of the Riobamba earthquake, not only were men engulfed, but animals, like mules, also sank into the fissures which were formed.

The fissures which were formed at the time of the Owen's Valley earthquake in 1872 extended for miles nearly parallel to the neighbouring Sierras. In some places the ground between the fissures sank twenty or thirty feet, and at one place about three miles east of Independence, a portion of the road was carried eighteen feet to the south by a fissure.[1]

Speaking generally, it may be said that all large earthquakes are accompanied by the formation of fissures. The Japanese have a saying that at the time of a large earthquake persons must run to a bamboo grove.

The object of this is to escape the danger of being engulfed in fissures, the ground beneath a bamboo grove being so netted together with fine roots that it is almost impossible for it to be rent open.

Materials discharged from fissures.—Together with the opening of cracks in the earth it often has happened that water, mud, vapours, gases, and other materials, have been ejected.

At the time of the Mississippi earthquake water, mixed with sand and mud, was thrown out with such violence that it spurted above the tops of the highest trees. In Italy such phenomena have often been repeated.

From the fissures which were formed in 1692 at the time of the earthquakes in Sicily, water issued which in some instances was salt.[2]

By the Cachar earthquake (January 10, 1869) numerous fissures were formed parallel to the banks of a river, from this water and mud were ejected. Dr. Oldham, who describes this earthquake, says that the

[1] *Am. Jour. Sci.* vol. iv.   [2] *Phil. Trans.* vol. xviii.

first shot of dry mud or sand was mistaken for smoke or steam.   The water was foul, and hotter than surface water at the time, but only slightly so ; and the sulphurous smell was nothing more than you would perceive in stirring up the mud at the bottom of any stagnant pool which had lain undisturbed for some time.[1]

In 1755, when Tauris was destroyed, boiling water issued from the cracks which were formed.   Similar phenomena were witnessed at a place eight miles from La Banca in Mexico, in the year 1820.   Part of this hot water was pure and part was muddy.

Sometimes the water which has been ejected has been so muddy that the mud has been collected to form small hills.   This was the case at the time of the Riobamba earthquake.   The mud in this case consisted partly of coal, fragments of augite, and shells of infusoria.

At the time of the Jamaica earthquake men who had fallen into crevices were in some cases thrown out again by issuing water.

Sometimes, as has already been mentioned, vapour, gases, and even flames issue from fissures.   Vapour of sulphur appears to be exceedingly common.   Kluge says that many fish were killed in consequence of the sulphurous vapours which rose in the sea near to the coast of New Zealand in 1855.

On December 14, 1797, an insupportable smell of sulphur was observed to have accompanied the earthquake which at that time shook Cumana, which was greatest when the disturbance was greatest.

Sulphurous fumes which were combustible were belched out of the earth at the time of the Jamaica earthquake in 1692.   The smell which accompanied this

[1] Oldham and Mallet, 'Cachar Earthquake,' *Proc. Geolog. Soc.* 1872.

was so powerful that it caused a general sickness which swept away about 3,000 persons.[1]

From the fissures formed at Concepcion in 1835, water, which was black and fœtid, issued.[2]

The earthquakes of New England in 1727 were accompanied by the formation of fissures, from which sand and water boiled out in sufficient quantity to form a quagmire. In some places ash and sulphur are said to have been ejected.

At one house the stink of sulphur accompanying the earthquake was so great that the family could not bear to remain in doors.[3]

Emanations of gas sometimes appear to have burst out from submarine sources.

Thus the earthquake at Lima, in March, 1865, was accompanied with great agitation of the water and an odour of sulphuretted and carburetted hydrogen. This former gas was developed to such an extent that the white paint of the U.S. ship 'Lancaster' was blackened.[4] With the smell, flames have sometimes been observed, as, for instance, at the time of the Lisbon earthquake.

At the time of the earthquakes of 1811 and 1813, in the Mississippi valley, steam and smoke issued from some of the fissures which were formed.

Instances are recorded where stones have been shot up from fissures unaccompanied by water, as, for instance, at the earthquake of Pasto (January, 1834). It is imagined that the propelling power must have been the sudden expansion of escaping gases.

It has been suggested that flames seen above fissures

[1] *Phil. Trans.* vols. li. and xviii. ; *Gent. Mag.* vol. xx. 212.
[2] *Trans. Royal Geog. Soc.* vol. vi.
[3] *Phil. Trans.* vols. xxxvi. and xxxix.
[4] *Am. Jour. of Sci.* 1865, vol. xl. p. 365.

might perhaps be due to the burning of materials like sulphur. Mr. D. Forbes, who examined the effects of the earthquakes of Mendoza, which were felt for a distance of 1,200 miles, says that where the hard rock came to the surface there were no traces of fissures, these being entirely confined to the alluvium. The rumours of fire and smoke having appeared at some of the fissures were without foundation, the presumed smoke being nothing but dust.[1]

In addition to flames lights appear often to have been observed, the origin of which cannot be easily explained.

The earthquake of November 22, 1751, at Genoa is said to have been accompanied by a light like that of a prodigious fire which seemed to arise out of the ground.[2]

Explanation of fissure phenomena.—The manner in which fissures are formed has already been explained when referring to the want of support in the face of hills (page 136).

Similar remarks may be applied to the banks of rivers and all depressions, whether natural or artificial, which have a steep slope. At such places the wave of shock emerges on a free surface, which, being unsupported in the direction of its motion, tends to tear itself away from the material behind, and form a fissure parallel to the face of the free surface. The distance of the fissure from the face of the free surface will, theoretically, be equal to half the amplitude of the wave of motion, one half tending to move forwards, and the other half backwards. The reason that water and other materials rush forth from fissures has been explained by Schüler as being due to cracks having been opened through impervious

---

[1] *Proc. Geolog. Soc. Ap.* 1875, p. 270.
[2] *Gent. Mag.* vol. xxi. p. 569.

strata, which, before the earthquake, by their continuity prevented the rising of subterranean water under hydro-static pressure.[1]

Kluge explains the coming up of the waters as being due to the same causes which he considers may be the origin of disturbances in the sea.

The most reasonable explanations of the eruption of water, mud, sand, and gas through fissures are those given by Oldham and Mallet in their account of the Cachar earthquake.

In the case of a horizontal shock passing through a bed of ooze or water-bearing strata, the elastic wave will tend to pack up the water during the forward motion to such an extent that it will flow or spout up through any aperture communicating with the surface. By the re-petition of these movements causing ejections, sand or mud cones, like those produced by a volcanic eruption, may be formed, and by a similar action water may be shot violently up out of wells, as was the case in Jamaica in 1692.

If an emergent wave acts through a water-bearing bed upon a superincumbent layer of impervious material, this upper layer is, during the upward motion, by its inertia suddenly pressed down upon the latter.

This pressure is equal to that which would raise the upper layer to a height equal to the amplitude of the motion of an earth particle, and with a velocity at least equal to the mean velocity of the earth particle resolved in the vertical direction.

For a moment the water bearing strata receive an enormous squeeze, and the water or mud starts up through any crevice which may be formed leading to the surface.

[1] *Jahrb. f. Min.* 1840, p. 173.

From this we see that liquids may rise far beyond the level due to hydrostatic pressure.[1]

Volger has attributed the origin of lights or flames appearing above fissures to the friction which must take place between various rocky materials at the time when the fissures are opened. As confirmatory of this he refers to instances where similar phenomena have been observed at the time of landslips. At the time of these landslips the heat developed by friction has been sufficiently intense to convert water into steam, the tension of which threw mud and earth into the air like the explosion of a mine.[2]

The gas eruptions which occasionally take place with earthquakes are probably due to the opening of fissures communicating with reservoirs or strata charged with products of natural distillation, or chemical action, which previously had accumulated beneath impervious strata. Of the existence of such gases we have abundant evidence. In coal mines we have fire-damp which escapes in increased quantities with a lowering of the barometrical pressure. In volcanic regions we have many examples of natural springs of carbon dioxide.

These various gases sometimes escape in quantity, or erupt without the occurrence of earthquakes. Rossi mentions an instance where a few years ago quantities of fish were killed by the eruption of gas in the Tiber, near Rome. Another instance is one which occurred at Follonica on April 6, 1874. On the morning of that day many of the streets and roads were covered with the dead bodies of rats and mice. It seemed as if it had rained rats. From the facts that the bodies of the creatures seemed healthy,

[1] Oldham and Mallet, ' Cachar Earthquake,' *Trans. Geolog. Soc. Ap* 1872.

[2] O. Volger, *Unters. üb. d. Phän. d. Erdb.* vol. iii. p. 414.

that the destruction had happened suddenly, and not
come on gradually like an epidemic, it was supposed that
they had been destroyed by an emanation of carbon dioxide. The fact that many of them lay in long lines
suggested the idea that they had been endeavouring to
escape at the time of the eruption.[1] If we can suppose
sudden developments of gas like this to have occasionally
accompanied earthquakes, we may sometimes have the
means of accounting for the sickness which has been felt.

Disturbances in lakes.—It has often been observed
that, at the time of large earthquakes, lakes have been
thrown into violent agitation, and their waters have been
raised or lowered. At the time of the great Lisbon earthquake, not only were the waters of European lakes thrown
into a state of oscillation, but similar effects were produced in the great lakes of North America. In some
instances, as in the case of small ponds, these movements
may be produced by the horizontal backward and forward
motion of the ground. , At other times they are probably
due to an actual tipping of a portion of their basins.
Movements like these latter will be again referred to in
the chapter on Earth Pulsations. On January 27, 1856,
there was a shock of earthquake at Bailyborough, Ireland,
which occasioned an adjacent lough to overflow its banks
and rush into the town with great impetuosity. In return-
ing it swept away two men, leaving behind a great quantity
of pike and eels of a prodigious growth.[2]

Disturbances in rivers.—Just as lakes have been
disturbed, so also have there been sudden disturbances in
rivers. Sometimes these have overflowed their banks,
whilst at other times they have been suddenly dried up.
In certain cases the reason that a portion of a river should

[1] *Meteorologia Endogena*, vol. i. p. 166.
[2] *Gent. Mag.* vol. xxvi. p. 91.

have become dry has been very apparent, as, for instance, at the time of the Zenkoji earthquake in Japan in 1847, when the Shikuma-gawa became partly dry in consequence of the large masses of earth which had been shaken down from overhanging cliffs damming a portion of its course, and thus forming, first, lakes, and subsequently, new water-courses. As another example, out of the many which might be quoted, may be mentioned the sudden drying up of the river Aboat, a tributary of the Magat, in the Philippine Islands, on July 27, 1881, shortly after a severe shock of earthquake. The water of this river ceased to flow for two hours, after which it reappeared with considerable increase of volume and of a reddish colour. Signor E. A. Casariego, who describes this, remarks that the phenomenon could easily be explained through the slipping down of the steep banks in narrow parts of its upper valley, by which means its flow had been obstructed until the water had time to accumulate and pass over or demolish the obstruction.

After the earthquake of Belluno (June 29, 1873), the torrent Tesa, which is ordinarily limpid, became very muddy.[1] Similar phenomena have been observed even in Britain, as, for instance, in 1787, when, at the time of a shock which was felt in Glasgow, there was a temporary stoppage in the waters of the Clyde. Again, in 1110, there was a dreadful earthquake at Shrewsbury and Nottingham, and the Trent became so low at Nottingham that people walked over it.

The earthquake of 1158, which was felt in many parts of England, was accompanied by the drying up of the Thames, which was so low that it could be crossed on foot even at London.[2]

---

[1] *Compte Rendu,* 1873, p. 66.
[2] *An Historical Account of Earthquakes,* p. 46.

Facts analogous to these are mentioned in the accounts of many large earthquakes. Sometimes rivers only become muddy or change their colour. In an account of the Lisbon earthquake we read that some of the rivers near Neufchâtel suddenly became muddy.[1]

At other times large waves are formed. Thus the earthquake of Kansas (April 24, 1867) apparently created a disturbance in the rivers at Manhattan, which rolled in a heavy wave from the north to the south bank.[2]

Sometimes curious phenomena have happened with regard to rivers without the occurrence of earthquakes. Thus, for instance, on November 27, 1838, there was a simultaneous stoppage of the Teviot, Clyde, and Nith.

In these rivers similar phenomena have been observed in previous years.

Again, on January 1, 1755, there was a sudden sinking of the river Frooyd, near Pontypool. This appears to have been due to the water sinking into chasms which were suddenly opened.[3]

Effects produced in springs, wells, fumaroles, &c.— Springs also are often affected by earthquakes. Sometimes the character of their waters change; those which were pure become muddy, whilst those which were hot have their temperature altered.

Sometimes springs have been dried up, whilst at other times new springs have been formed.

This latter was the case in New England (October 27, 1727). In some places springs were formed, whilst at other places they were either entirely or partly dried up.[4]

At and near Lisbon, in 1755, some fountains became

<hr>

[1] *Phil. Trans.* vol. xlix. p. 436.   [2] *Am. Jour. Sci.* vol. xlv. p. 129.
[3] *Phil. Trans.* vol. xlix. p. 547.   [4] *Ibid.* vols. xlii. and xxxix.

muddy, others decreased, others increased, and others dried up. At Montreux, Aigle, and other places, springs became turbid.

The baths at Toplitz, in Bohemia, which were discovered in A.D. 762, were seriously affected by the same earthquake. Previous to the earthquake it is said that they had always given a constant supply of hot water. At this time, however, the chief spring sent up vast quantities of water and ran over. One hour before this it had grown turbid and flowed muddy. After this it stopped for about one minute, but recommenced to flow with prodigious violence, driving before it considerable quantities of reddish ochre. Finally, it settled back to its original clear state and flowed as before.[1]

In 1855, at the earthquake of Wallis, many new springs burst forth, and some of these in Nicolai Thale were so rich in iron that they quickly formed a deposit of ochre.

At the time of the Belluno earthquake (June 29, 1873), a hot spring, La Vena d'Oro, suddenly became red.[2]

The following examples of like changes are taken from the writings of Fuchs.[3]

In 1738 the hot springs of St. Euphema rose considerably in their temperature.

During the earthquake of October, 1848, the hot springs of Ardebil, which usually had a temperature of from 44° to 46° C., rose so high that their temperature was sufficient to cause scalding.

At the time of the earthquake of Wallis, in 1855, the temperature of hot springs rose 7°, and the quantity of water increased three times.

During the earthquake of 1835 in Chili, the springs

<hr />

[1] *Phil. Trans.* vol. xlix. part i.
[2] *Compte Rendu,* 1873, part ii. p. 66.
[3] *Die Vulcan. Ers. d. Erde,* C. W. C. Fuchs.

of Cauquenes fell from 118° to 92° F. Subsequently, however, they again rose.

Fumaroles are similarly disturbed. Thus, at the time of the earthquakes of Martinique (September, 1875), the fumaroles there showed an abnormal activity.[1]

Wells often appear to be acted upon in the same manner as springs.

At the time of the California earthquake (April, 1855), the level of the water in certain wells was raised ten to twelve feet.

A consequence of the earthquake at Neufchâtel, in 1749, was to fill some of the wells with mud.[2] At Constantinople, on September 2, 1754, wells became dry.[3]

Explanation of the above phenomena.—That the water in springs and wells should be caused to rise at the time of an earthquake, admits of explanation on the supposition of compressions taking place similar to those which cause the rise of water in fissures. That the water in wells and springs should be rendered turbid, is partly explained on the supposition of more or less dislocation taking place in the earthy or rocky cavities in which they are contained or through which they flow.

At the time of a large earthquake it is extremely probable that there is a general disturbance in the lines of circulation of subterranean waters and gases throughout the shaken area. By these disturbances, new waters may be brought to the surface, two or more lines of circulation may be united, and the flow of a spring or supply of a well be augmented. Fissures, through which waters reached the surface, may be closed, wells may become dry, or springs may cease to flow, hot springs may have their

---

[1] *Compte Rendu,* 1875, p. 693.   [2] *Gent. Mag.* vol. xix. p. 190.
[3] *Phil. Trans.* vol. xlix. p. 115.

temperature lowered by the additions of cold water from another source, and, in a similar manner, waters may be altered in their mineralisation.   An important point to be remembered in this consideration is the mutual dependence of various underground water supplies, and the area over which any given supply may circulate.   A well in the higher part of Lincoln Heath is said to be governed by the river Trent, which is ten miles distant ; when the river rises the well rises in proportion, and when the river falls the water in the well falls.[1]

The change which is usually observed in hot springs is, that before or with earthquakes they increase in temperature, but afterwards sink back to their normal state. This increase in temperature may possibly be due to communication being opened with new or deeper centres of volcanic activity, or a temporarily increased rate of flow.

That the water issuing from newly formed fissures or springs should be hot, might be explained on the supposition of its arising from a considerable depth, or from some volcanic centre.   It might also be attributed to the heat developed by friction at the opening of the fissures. These changes which earthquakes produce upon the underground circulation of waters are phenomena deserving especial attention.   Although we know much about the circulation of surface water, it is but little that we yet know about the movement of the streams hidden from view, from which these surface waters have their sources. Earthquakes may be regarded as gigantic experiments on the circulatory system of the earth, which, if properly interpreted, may yield information of scientific and utilitarian value.

The sudden elevations, depressions, or lateral shifting of large tracts of country at the time of destructive earth-

[1] *Gent. Mag.* vol. xxi. 1751.

quakes are phenomena with which all students of geology are familiar. In most cases these displacements have been permanent; and evidences of many of the movements which occurred within the memory of man, remain as witnesses of the terrible convulsions with which they were accompanied.

Movements on coast lines and level tracts.—At the time of the great earthquake of Concepcion, on February 20, 1835, much of the neighbouring coast line was suddenly elevated four or five feet above sea level. This, however, subsequently sank until it was only two feet. A rocky flat, off the island of Santa Maria, was lifted above high-water mark, and left covered with ' gaping and putrefying mussel-shells, still attached to the bed on which they had lived.' The northern end of the island itself was raised ten feet and the southern extremity eight feet.[1]

By the earthquake of 1839, the island of Lemus, in the Chonos Archipelago, was suddenly elevated eight feet.[2]

Of movements like these, especially along the western shores of South America, Darwin, who paid so much attention to this subject, has given many examples. In 1822, the shore near Valparaiso was suddenly lifted up, and Darwin tells us that he heard it confidently asserted ' that a sentinel on duty, immediately after the shock, saw a part of a fort which previously was not within the line of his vision, and this would indicate that the uplifting was not vertical.'[3] That the large areas of land should be shifted permanently in horizontal directions, as well as vertically, we should anticipate from the observations which we are able to make upon large fissures which are caused by earthquakes.

Another remarkable example of sudden movement in

---

[1] *Jour. Royal Geo. Soc.* vol. vi. p. 319.
[2] Darwin, *Geolog. Observations*, p. 232.   [3] *Ibid.* p. 245.

the rocky crust is that which took place during the earthquakes of 1811–12 in the valley of the Mississippi, near to
the mouth of the Ohio, which was convulsed to such a
degree, that lakes, twenty miles in extent, were formed in
the course of an hour.   This country, which is called the
' sunk country,' extends some seventy to eighty miles
north and south, and thirty miles east and west.[1]

In the ' Gentleman's Magazine ' we read of the little
territory of Causa Nova, in Calabria, being sunk twenty-
nine feet into the earth by an earthquake, without throwing
down a house.   The inhabitants, being warned by a noise,
escaped into the fields, and only five were killed.[2]

Other examples of these permanent dislocations of
strata are to be found in almost every text-book on
geology.

Geological changes produced. — Passing over the
accounts of earth movements which are more or less ficti-
tious, and confining our attention to the well authenticated
facts, we see at once the important part which earthquakes
have played as agents working geological changes.   Even
in the nineteenth century long tracts of coast, as in Chili
and New Zealand, have been raised, whilst other areas, like
the Delta of the Indus, have been sunk.   Sir H. Bartle
Frere, speaking about the disturbance which took place in
this latter region in 1819, remarks that all the canals
drawn from the Fullalee River ceased to run for about
three days, probably indicating a general upheaval of the
lower part of the canal.   In consequence of the earthquakes in former times it is not unlikely that water-courses
have ceased to flow, water has decreased in wells, and
districts have been depopulated.[3]

---

[1] Lyell, *Principles of Geology*, vol. ii. pp. 107–8.
[2] *Gent. Mag.* 1733, vol. iii. p. 217.
[3] ' Earthquakes of Cutch,' *Jour. Royal Geo. Soc.* vol. xl.

M

Sometimes these changes have taken place gradually, and sometimes with violence. Mountains have been toppled over, valleys have been filled, cities have been submerged or buried.

With the records of these convulsions before us, we see that seismic energy yet exhibits a terrible activity in changing the features of the globe.

Reason of these movements.—To formulate a single reason for these catastrophes would be difficult. Where they are of the nature of landslips, or materials have been dislodged from mountain sides, the cause is evidently the sudden movement of this ground acting upon strata not held together in a sufficiently stable condition. A similar explanation may be given for the sudden elevations or depressions of strata in a district removed from the centre where the disturbance had its origin. The seismic effort exhibits itself in a certain area round its origin as a sudden push, and by this push, strata are fractured and caused to move relatively to each other.

At or near to the origin of an earthquake it might be argued that it was the sudden falling of rocky strata towards a position of stable equilibrium that caused the shaking, and in such a case the movements referred to may be regarded as the cause rather than the effect of an earthquake.

A subject closely connected with the sudden dislocation of strata, is the production of secondary or consequent earthquakes, due to the disturbance of ground in a critical state (see p. 248).

CHAPTER IX.

DISTURBANCES IN THE OCEAN.

Sea vibrations—Cause of vibratory blows—Sea waves: Preceding earthquakes; Succeeding earthquakes — Magnitude of waves— Waves as recorded in countries distant from the origin—Records on tide gauges—Waves without earthquakes—Cause of waves— Phenomena difficult of explanation—Velocity of propagation— Depth of the ocean—Examples of calculations—Comparison of velocities of earthquake waves with velocities which ought to exist from the known depth of the ocean.

Sea vibrations.—Whilst residing in Japan I have had many opportunities of conversing with persons who had experienced earthquakes when on board ships, and it has often happened that these same earthquakes have been recorded on the shore. For example, at the time of every moderately severe earthquake which has shaken Yokohama, the same disturbance has been felt on board the ships lying in the adjoining harbour. In some cases the effect had been as if the ship was grounding; in others, as if a number of sharp jerks were being given to the cable. The effect produced upon a man-of-war lying in the Yokohama harbour on the evening of March 11, 1881, was described to me as a 'violent irresistible shaking.' Vessels eighty miles at sea have recorded and timed shocks which were felt like sudden blows. These were accompanied by a noise described as a 'dull rattle like thunder.'

M 2

In none of the cases here quoted was any disturbance of the water observed.

The great earthquake of Lisbon was felt by vessels on the Atlantic, fifty miles away from shore.

On February 10, 1716, the vessels in the harbour of New Pisco were so violently shaken that both ropes and masts were broken, and yet no motion in the water was observed. Some have described these shocks like those which would be produced by the sudden dropping of large masses of ballast in the hold of the vessel. Other cases are known where rigging was damaged, and even cannon have been jerked up and down from the decks on which they rested.

Cause of vibratory blows.—From the rattling sound which has accompanied some of these submarine shocks, many of which, it may be remarked, have never been recorded as earthquakes upon neighbouring shores, it does not seem improbable that they may have been the result of the sudden condensation of volumes of steam produced by submarine volcanic eruptions.

As confirmatory of this supposition we have the fact, that many of the marine disturbances which might be called 'sea-quakes,' have been observed in places which are close to, or in the line of, volcanic vents. Thus, M. Daussy, who has paid special attention to this subject, has collected evidence to show that a large number of shocks have been felt by vessels in that portion of the Atlantic between Cape Palamas, on the west coast of Africa, and Cape St. Roque, on the east coast of South America.[1]

Some of the vessels only felt shocks and tremblings, but others saw smoke, and some even collected floating ashes. In considering the submarine shocks of this par-

[1] M. Daussy, ' Sur l'existence probable d'un volcan sousmarin situé par environ 0° 20′ de lat. S., et 22° 0′ de long. ouest,' *Comptes Rendus*, vol. vi. p. 512.

ticular area, we must bear in mind that it lies in the line of Iceland, the west coast of Scotland, the Azores, Canaries, Cape de Verd Islands, St. Helena, and other places, all of which, if not at present in volcanic activity, shew evidence of having been so within recent times. The connection between volcanic action and earthquakes will be again referred to.

Sea waves.—Although in the above-mentioned instances sea waves have not been noticed, it is by no means uncommon to find that destructive earthquakes have been accompanied by waves of an enormous size, which, if the earthquake has originated beneath the sea, have, subsequently to the shaking, rolled in upon the land, to create more devastation than the actual earthquake. It may, however, be mentioned that a few exceptional cases exist when it is said that the sea wave has preceded the earthquake, as, for example, at Smyrna, on September 8, 1852.

Again, at the earthquake in St. Thomas, in 1868, it is said that the water receded shortly *before* the first shock. When it returned, after the second shock, it was sufficient to throw the U.S. ship ' Monagahela ' high and dry.[1]

Another American ship, the ' Wateree,' was also lost in 1868 by being swept a quarter of a mile inland by the sea wave which inundated Arequippa.

Much of the great destruction which occurred at the time of the great Lisbon earthquake was due to a series of great sea waves, thirty to sixty feet higher than the highest tide, which swamped the town. These came in about an hour after the town had been shattered by the motion of the ground.

The first motion in the waters was their withdrawal, which was sufficient to completely uncover the bar at the mouth of the Tagus. At Cadiz, the first wave, which was

[1]. *Am. Jour. Sci.* vol. xlv. p. 133.

the greatest, is said to have been sixty feet in height.
Fortunately the devastating effect which this would have
produced was partially warded off by cliffs.

At the time of the Jamaica earthquake (1692) the sea
drew back for a distance of a mile.

In South America sea waves are common accompani-
ments of large earthquakes, and they are regarded with
more fear than the actual earthquakes.

On October 28, 1724, Lima was destroyed, and on
the evening of that day the sea rose in a wave eighty feet
over Callao. Out of twenty-three ships in the harbour,
nineteen were sunk, and four others were carried far
inland. The first movement which is usually observed is
a drawing back of the waters, and this is so well known
to precede the inrush of large waves, that many of the
inhabitants in South America have used it as a timely
warning to escape towards the hills, and save themselves
from the terrible reaction which, on more than one occasion,
has so quickly followed.

At Caldera, near to Copiapo, on May 9, 1877, which was
the time when Iquique was devastated, the first motion
which was observed in the sea was that it silently drew
back for over 200 feet, after which it rose as a wave over
five feet high. At some places the water came in as
waves from twenty to eighty feet in height.

At Talcahuano, on the coast of Chili, in 1835, there was
a repetition of the phenomena which accompanied the
destruction of Penco in 1730 and 1751. About forty
minutes after the first shock, the sea suddenly retired.
Soon afterwards, however, it returned in a wave twenty
feet high, the reflex of which swept everything towards
the sea. These phenomena were repeated three times.[1]

When Callao and Lima were destroyed, in 1746, the

[1] *Am. Jour. Sci.* vol. xiv. p. 209.

sea first drew back, then came in as waves, four or five minutes after the earthquake. Altogether, between October 28 and February 24, 451 shocks were counted. At one time the sea came in eighty feet above its usual level. One account says that the large waves came in forty-one and a half hours after the first shock, and seventeen and a half hours after comparative tranquillity had prevailed.[1]

At the eruption of Monte Nuovo, near Naples, in September 27, 1538, the water drew back forty feet, so that the whole gulf of Baja became dry.

In 1696, at the time of the Catanian earthquake, the sea is said to have gone back 2,000 fathoms. Instances are recorded where the sea has receded several miles.

The time taken for the flowing back of the sea is usually very different. Sometimes it has only been five or six minutes, whilst at other times over half an hour, and there are records where the time is said to have been still longer.

Thus, at the earthquake of Santa (June 17, 1678), the sea is stated to have gone as far back as the eye could reach, and did not rise again for twenty-four hours, when it flooded everything.

In 1690 at Pisco the sea went back two miles, and did not return for three hours. When it returns it does so with violence, and examples of the heights to which it may reach have been given. The greatest sea wave yet recorded, according to Fuchs, is one which, on October 6, 1737, broke on the coast of Lupatka, 210 feet in height.

There are, however, cases known where the sea has returned as gradually as it went out. Thus, on December

---

[1] D. C. F. Winslow, ' Tides at Tahiti,' *Am. Jour. Sci.* 1865, p. 45 ; also Mallet's *Catalogue of Earthquakes.*

4, 1854, when Acapulco was destroyed, the sea is said to have returned as gently as it went out.

When sea waves have travelled long distances from their origin, as, for instance, whenever a South American wave crosses the Pacific to Japan, the phenomena which are observed are like those which were observed at Acapulco; the sea falls and rises, at intervals of from ten minutes to half an hour, to heights of from six to ten feet, without the slightest appearance of a wave. Its phenomenon is like that of an unusually high tide, which repeats itself several times per hour. Even if we watch distant rocks with a telescope, although the surface of the ocean may be as smooth as the surface of a mirror, there is not the slightest visible evidence of what is popularly called a wave. The sea being once set in motion it continues to move as waves of oscillation for a considerable time. In 1877, as observed in Japan, the motion continued for nearly a whole day. The period and amplitude of the rise and fall were variable, usually it quickly reached a maximum, and then died out gradually. As observed in a self-recording tide gauge at San Francisco, the disturbance lasted for about four days. A diagram of this is here given. In its general appearance it is very similar to the records of other earthquake waves. The large waves represent the usual six hours rise and fall of the tides; usually these are fairly smooth curves. Superimposed on the large waves are the smaller zigzag curves of the earthquake disturbance, lasting with greater or less intensity for several days. As these curves are drawn to scale—horizontally for hours, and vertically one fifth inch to the foot, to show the extent of the rise and fall—they will be easily understood.

Sometimes, as in the present example, the first movement in the waters is that of an incoming wave. In many instances, however, this observation may be due to the slow and more gentle phenomena of the previous drawing out of the water, which in a steep waste, or when the water is rough, would be difficult to observe, not having been remarked.

The distance to which these sea waves have extended has usually been exceedingly great.

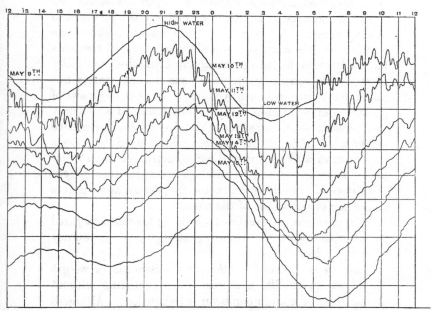

FIG. 28.—Record of Tide Gauge at Port Point, San Francisco. Showing Earthquake Waves of May 1877.

The sea wave of the Iquique earthquake of May 9, 1877, like many of its predecessors, was felt across the basin of the whole Pacific, from New Zealand in the south, to Japan and Kamschatka in the north. And but for the intervention of the Eurasian and American continents would have made itself appreciable over the surface of the whole of our globe. At places on the South American coast, it has been stated that the height of the waves varied from twenty to eighty feet. At the Samoa Islands the heights varied from six to twelve feet. In New Zealand the sea rose and fell from three to twenty feet. In Australia the heights to which the water oscillated were similar to those observed in New Zealand. In Japan it rose and fell from five to ten feet. In this latter country, the phenomena of sea waves which follow a destructive earthquake on the South American coast are so well known that old residents have written to the local papers announcing the probability of such occurrences having taken place some twenty-five hours previously in South America. In this way news of great calamities has been anticipated, details of which only arrived some weeks subsequently. Just as the destructive earthquakes of South America have announced themselves in Japan, in a like manner the destructive earthquakes of Japan have announced themselves upon the tide gauges of California. Similarly, but not so frequently, disturbances shake the other oceans of the world.

For example, the great earthquake of Lisbon propagated waves to the coasts of America, taking on their journey nine and a half hours.

Sea waves without earthquakes.—Sometimes we get great sea waves like abnormal tides occurring without any account of contemporaneous earthquakes. Although earthquakes have not been recorded, these ill understood phenomena are usually attributed to such movements.

Several examples of these are given by Mallet. Thus, at 10 A.M. on March 2, 1856, the sea rose and fell for a considerable distance at many places on the coast of Yorkshire. At Whitby, the tide was described ebbing and flowing six times per hour, and this to such a distance

that a vessel entering the harbour was alternately afloat
and aground.

In 1761, on July 17, a similar phenomena was observed
at the same place.

A like occurrence took place at Kilmore, in the county
of Wexford, on September 16, 1864, when the water ebbed
and flowed seven times in the course of two hours and a
half. These tides, which appear to have taken about five
minutes to rise and five minutes to fall, were seen by an
observer approaching from the west as six distinct ridges
of water. The general character of the phenomena appears
to have been very similar to that which was produced at
the same place by the Lisbon earthquake of 1755 ; and
the opinion of those who saw and wrote about their
occurrence was that it was due to an earthquake disturbance. Such phenomena are not uncommon on the Wexford
coast, where they are popularly known as ' death waves,'
probably in consequence of the lives which have been lost
by these sudden inundations.

They have also been observed in other parts of Ireland,
the north-east coast of England, and in many parts of the
globe. They will be again referred to under the head of
earth pulsations.

Cause of sea waves.—Mallet, who in his report to the
British Association in 1858, writes upon this last-mentioned
occurrence at considerable length, whilst admitting that
many may have originated from earthquakes, he thinks it
scarcely probable that an earthquake blow, sufficiently
powerful to have produced waves like those observed at
Kilmore, should not have been felt generally throughout
the south of Ireland. He, therefore, suggests that sometimes waves like the above might be produced by an
underwater slippage of the material forming the face of a
submarine bank, the slope of which by degradation and

deposition, produced by currents, had reached an angle beyond the limits of repose of the material of which it was formed. Mallet does not insist upon the existence of these submarine landslips, but only suggests their existence as a means of explaining certain abnormal sea waves which do not appear to have been accompanied by earthquakes.

In the generality of cases sea waves are accompanied by earthquakes, but it may often happen that the connection between the two is difficult to clearly establish. One simple explanation for the origin of waves occurring with earthquakes, is, that in consequence of the earthquake a large volume of water suddenly finds its way into cavities which have been opened, the disturbance produced by the inrush giving rise to waves.

A second explanation is, that the land along a shore is caused by an earthquake to oscillate upwards, the water running off to regain its level. A supposition like this is negatived by the fact that these disturbances are felt far away from the chief disturbance, on small islands. Also, it may be added, that the whole disturbance appears to approach the land from the sea, and not in the opposite direction. Thus, in the earthquake of Oahu (February 18, 1871), it was remarked that the shock was first felt by the ships farthest from the land.[1]

Another suggestion is that the waves are due to a sudden heaving up of the bottom of the ocean. If this lifting took place slowly, then the first result would be that the water situated over the centres of disturbance would flow away radially in all directions from above the area of disturbance.

If, however, the submarine upheaval took place with great rapidity, say by the sudden evolution of a large

[1] *Am. Jour. Sci.* vol. i. p. 469.

volume of steam developed by the entry of water into a volcanic vent, as the water was heaped above the disturbed area, water might run in radially towards this spot.

Supposing a primary wave to be formed in the ocean by any such causes, then the falling of this will cause a second wave to be formed, existing as a ring round the first one. The combined action of the first and second wave will form a third one, and so the disturbance, starting from a point, will radiate in broadening circles. During the up and down motion of these waves, the energy which is imparted to any particle of water will, on account of the work which it has to do in displacing its neighbours, by frictional resistance, gradually grow less and less, until it finally dies away. The waves which are the result of this motion will also grow less and less.

If a series of sea waves were produced by a single disturbance, we see that these will be of unequal magnitude. Now, for small waves, the velocity with which they travel depends upon the square root of their lengths; but with large waves, like earthquake waves, the velocity depends upon the square root of the depth of water, and these latter travel more quickly than the former.

If, therefore, we have a series of disturbances of unequal magnitude producing sea waves, which, from the series of shocks which have been felt upon shores subsequently invaded by waves, seems in all probability often to have been the case, it is not unlikely that the waves of an early disturbance may be overtaken and interfered with by a series which followed.

These considerations help us to understand the appearance of the records on our tide gauges, and also the phenomena observed by those who have recorded tidal waves as they swept inwards upon the land. For instance, we understand the reason why sea waves, as observed at

places at different distances from the origin of a disturbance, should be of different heights. We also see an
explanation for the fact that small waves should sometimes appear to be interpolated between large ones, and
that these should occur at varying intervals.

The fact that whenever a wave is produced, a certain
quantity of water must be drawn from the level which
surrounds it, in order that it should be formed, explains
the phenomena that the sea is often observed first to
draw back. Out in the open ocean it is drawn from the
hollow between two waves. As has been pointed out by
Darwin, it is like the drawing of the water from the shore
of a river by a passing steamer.

The difference in the height of waves, as observed at
places lying close to each other, is probably due to the configuration of the coast, the interference of outlying islands,
reefs, &c.—causes which would produce similar effects in
the height of tide.

As a wave approaches shallow water it gradually increases in height, its front slope becomes steep, and its
rear slope gentle, until finally it topples over and breaks.
This increasing in height of waves is no doubt connected
with the destruction of Talcahuano and Callao, which are
situated at the head of shallow bays. Valparaiso, which is
on the edge of deep water, has never been overwhelmed.[1]
Another case tending to produce anomalies in the character of waves would be their reflection and mutual interference, the reflections due to the configuration of the
ocean bed and coast lines.

The complete phenomena which may accompany a
violent submarine disturbance are as follows :—

By the initial impulse of explosion or lifting of the
ground, a 'great sea wave' is generated, which travels

---

[1] Darwin, *Researches in Geology, &c.*, p. 378.

shorewards with a velocity dependent upon its size and the depth of the ocean. At the same instant, a 'sound wave' may be produced in the air, which travels at a quicker rate than the 'great sea wave.' A third wave which is produced, is an 'earth wave,' which will reach the shore with a velocity dependent on the intensity of the impulse and the elasticity of the rocks through which it is propagated. This latter, which travels the fastest, may carry on its back a small 'forced sea wave.' On reaching the shore and passing inland, this 'earth wave' will cause a slight recession of the water as the 'forced sea wave' slips from its back.

As these 'forced sea waves' travel they will give blows to ships beneath which they may pass, being transmitted from the bottom of the ocean to the bottom of the ships like sound waves in water. At the time of small earthquakes, produced, for example, by the explosion of small quantities of water entering volcanic fissures, or by the sudden condensation of steam from such a fissure entering the ocean, aqueous sound waves are produced, which cause the rattling and vibrating jars so often noticed on board ships.

Phenomena difficult of explanation.—Although we can in this way explain the origin and phenomena of sea waves, we must remember, as Kluge has pointed out, that it is not the simple backward and forward movement of the ground which produces sea waves, and that the majority of earthquakes which have occurred in volcanic coasts have been unaccompanied by such phenomena. Out of 15,000 earthquakes observed on coast lines, only 124 were accompanied by sea waves.[1] Out of 1,098 earthquakes catalogued by Perrey for the west coast of South America, only nineteen are said to have been accompanied

[1] Kluge, *Jahrb. f. Min.* 1861, p. 977.

by movements in the waters. According to the 'Geographical Magazine' (August 1877, p. 207), it would seem that out of seventy-one severe earthquakes which have occurred since the year 1500 upon the South American coast many have been accompanied by sea waves. Darwin also remarks, when speaking of South America, that almost every large earthquake has been accompanied by considerable agitation in the neighbouring sea.[1]

On April 2, 1851, when many towns in Chili were destroyed, the sea was not disturbed. At the time of the great earthquake of New Zealand (June 23, 1855), although all the shocks came from the sea, yet there was no flood. The small shock of February 14, however, was accompanied by a motion in the sea.

To these examples, which have been chiefly drawn from the writings of Fuchs, must be added the fact that the greater number of disturbances which are felt in the north-eastern part of Japan, although they emanate from beneath the sea, do not produce any visible sea waves. They are, however, sufficient to cause a vibratory motion on board ships situated near their origin.

Another point referred to by Fuchs, as difficult of explanation, is, that the water, when it draws back, often does so with extreme slowness, and farther, in some instances, it has not returned to its original level. That the sea might be drawn back for a period of fifteen or thirty minutes is intelligible, when we consider the great length of the waves which are formed. Cases where it has retired for several hours or days, and when its original level is altered, appear only to be explicable on the assumption of more or less permanent changes in the levels of the ground. For example, in the earthquake of 1855 which shook New Zealand, the whole southern portion of the northern island was raised several feet.

[1] Darwin, *Voyage of a Naturalist*, p. 309.

These sudden alterations in the levels of coast lines have already been referred to.

Other points which are difficult to understand are the occurrence of disturbances in the sea at the time of feeble earthquakes, and with earthquakes occurring in distant places. As examples of such occurrences, Fuchs quotes the following : ' On May 16, 1850, at 4.28 A.M., an earthquake took place in Pesth, and at 7.30 a motion was observed in the sea at Livorno. Again, at the time of the earthquake of December 19, 1850, which shook Heliopolis, a flood suddenly came in upon Cherbourg.' May not these phenomena be the result of an earth pulsation, which produced an earthquake at one point, and a sea wave at another ?

Equally difficult to understand are the observations when the disturbance in the sea has occurred several hours after an earthquake ; as, for instance, at Batavia, in 1852, when there was an interval of two hours ; and to this must be added the observations where the motion of the sea has preceded that of the earthquake—as, for instance, in 1852, at Smyrna. Whilst recognising the fact that it is possible to suggest explanations for many of these anomalies, we must also bear in mind that they are, generally speaking, exceptional, and, in some instances, may possibly be due to errors in observations.

Velocity of propagation of sea waves, and depth of the ocean.—It has long been known to physical science that the velocity with which a given wave is propagated along a trough of uniform depth, holds a relation to the depth of the trough.

If v is the velocity of the wave, and h the depth of the trough, this relation may be expressed as follows :—

$$h = \frac{v^2}{g} \text{ or } h = \left(\frac{v}{k} \right)^2$$

N

Where $g = 32 \cdot 19$ and $k = 5 \cdot 671$.

It will be observed that these two formulæ (the first of which is known as Russell's formula, and the second as Airy's) are practically identical.

The apparent difference is in the average value assigned to the constant.

For large waves such as we have to deal with, it would be necessary, if we were desirous of great accuracy, to increase the value of h by some small fraction of itself. We might also make allowance for the different values of g, according to our position on the earth's surface. With these formulæ at our disposal it is an easy matter, after having determined the velocity with which a wave was propagated, to determine the average depth of the area over which it was transmitted.

In making certain earthquake investigations the reverse problem is sometimes useful—namely, determining the velocity with which a sea wave has advanced upon a shown line, from a knowledge of the depth of the water in which it has been propagated.

Calculations of the average depths of the Pacific, dependent on the velocity with which earthquake waves have been propagated, have been made by many investigators.

In most cases, however, in consequence of having assumed the wave to have originated on a coast line, when the evidence clearly showed it to have originated some distance out at sea, the calculations which have been made are open to criticism. The average depths which I obtained for various lines across the Pacific appear to be somewhat less than the average depth as given by actual soundings. We must, however, remember that the common error in actual soundings is that they are usually too great, it being difficult in deep-sea sounding to determine when the lead actually reaches the bottom. Until oceans have been more thoroughly surveyed with the improved forms of sounding apparatus, we shall not be able to verify the truth of the results which have been given to us by earthquake waves.

EXAMPLES OF CALCULATIONS ON SEA WAVES.

1. *The wave of* 1854.—This wave originated near Japan, and it was recorded on tide gauges at San Francisco, San Diego, and Astoria.

On December 23, at 9.15. A.M., a strong shock was felt at Simoda in Japan, which, at 10 o'clock, was followed by a large wave thirty feet in height. The rising and falling of the water continued until noon. Half an hour after, the movement became more violent than before. At 2.15 P.M. this agitation decreased, and at 3 P.M. it was comparatively slow. Altogether there were five large waves.

On December 23 and 25, unusual waves were recorded upon the self-registering tide gauges at San Francisco, San Diego, and Astoria.

At San Francisco three sets of waves were observed. The average time of oscillation of one of the first set was thirty-five minutes, whilst one of the second and third sets was almost thirty-one minutes.

At San Diego three series of waves were also shown, but with average times of oscillation of from four to two minutes shorter than the waves at San Francisco.

The San Francisco waves appear to indicate a recurrence of the same phenomena.

The record at San Diego shows what was probably the effect of a series of impulses, the heights increasing to the third wave, then diminishing, then once more renewed, after which it died away.

The result of calculations based on these data were :—

	Distance geographical miles	Time of transmission	Velocity in feet per sec.	Depth of ocean in fathoms
		h. m.		
Simoda to San Diego	4917	12 13	545	2100
Simoda to San Francisco	4527	12 39	528	2500 or 2230

The difference for the depths in the San Francisco path depends whether the length of the waves is reckoned at 210 or 217 miles. The length of the waves on the San Diego path were 186 or 192 miles.[1]

The wave of 1868.—On August 11, 1868, a sea wave ruined many cities on the South American coast, and 25,000 lives were lost. This wave, like all the others, travelled the length and breadth of the Pacific.

In Japan, at Hakodate, it was observed by Captain T. Blakiston, R.A., who very kindly gave me the following account :

On August 15, at 10.30 A.M., a series of bores or tidal waves commenced, and lasted until 3 P.M. In ten minutes there was a difference in the sea level of ten feet, the water rising above high water and falling below low water mark with great rapidity. The ordinary tide is only two and a half to three feet. The disturbance producing these waves originated between Iquique and Arica, in about lat. 18.28 S. at about 5 P.M. on August 13. In Greenwich time this would be about 13h. 9m. 40s. August 13. The arrival of the wave at Hakodate in Greenwich time would be about 14h. 7m. 6s. August 14: that is to say, the wave took about 24h. 57m. to travel about 8,700 miles, which

[1] Prof. A. D. Bache, *United States Coast Survey Report*, 1855, p. 342.

gives us an average rate of about 511 feet per second,
These waves were felt all over the Pacific. At the
Chatham Islands they rushed in with such violence that
whole settlements were destroyed. At the Sandwich
Islands the sea oscillated at intervals of ten minutes for
three days.

Comparing this wave with the one of 1877 we see that
one reached Hakodate with a velocity of 511 feet per second,
whilst the other travelled the same distance at 512 feet
per second.

An account of this earthquake wave has been given
by F. von Hochstetter (' Über das Erdbeben in Peru am 13·
August 1868 und die dadurch veranlassten Fluthwellen
im Pacifischen Ocean,' Sitzungsberichte der Kaiserl.
Akademie der Wissenschaften, Wien 58. Bd., 2. Abth.
1868). From an epitome of this paper given in ' Petermann's Geograph. Mittheil.' 1869, p. 222, I have drawn
up the following table of the more important results
obtained by F. von Hochstetter.

The wave is assumed to have originated near Arica.

	Distance sea miles from Arica	Time taken by wave	Velocity in feet per second	Depth of ocean in feet
		h. m.		
Valdivia . . .	1,420	5 0	479	7,140
Chatham Islands .	5,520	15 19	608	11,472
Lyttleton . . .	6,120	19 18	533	8,838
Newcastle . .	7,380	22 28	538	9,006
Apia (Samoa) . .	5,760	16 2	604	11,346
Rapa . . .	4,057	11 11	611	11,598
Hilo	5,400	14 25	555	9,568
Honolulu . . .	5,580	12 37	746	17,292

Calculations on the same disturbance are also given by
J. E. Hilgard.[1]

Assuming the origin of the wave to have been at Arica,
his results are as follows :

[1] *United States Coast Survey Report*, or *Am. Jour. Sci.* vi. p. 77.

	Distance from Arica	Time of transmission	Nautical miles per hour	Mean depth of ocean
	miles	h. m.		feet
San Diego . . .	4,030	10 55	369	12,100
Fort Point . .	4,480	12 56	348	10,800
Astoria . . .	5,000	18 51	265	6,200
Kodiak . . .	6,200	22 00	282	7,000
Rapa	4,057	10 54	372	12,200
Chatham Islands .	5,520	15 01	368	12,100
Hawaii . . .	5,460	14 10	385	13,200
Honolulu . . .	5,580	12 18	454	18,500
Samoa . . .	5,760	15 38	368	12,100
Lyttelton . . .	6,120	19 01	322	9,200
Newcastle . . .	9,380	22 10	332	9,800
Sydney . . .	7,440	23 41	314	8,800

The wave of 1877.—Two sets of calculations have been made upon the wave of 1877 by Dr. E. Geinitz of Rostock.[1]

The following table is taken from Dr. Geinitz's second paper, in which there are several modifications of his first results. The origin of the disturbance is assumed to have been near Iquique.

Observation stations	Distance from Iquique geol. miles	Arrival of wave	Time taken by wave	Velocity in feet per second	Mean depth of ocean in fathoms
		h. m.	h. m.		
Taiohāc (Marquesa Islands)	4,086	8 40 A.M.	12 15	563·8	1,647
Apia (Samoa) . . .	5,740	12 0 „	15 30	610·4	1,930
Hilo (Sandwich Islands) .	5,526	10 24 „	14 0	667·9	2,310
Kahuliu „ .	5,628	10 30 „	14 5	675·2	2,361
Honolulu „ .	5,712	10 50 „	14 25	669·7	2,319
Wellington (New Zealand) .	5,657	2 40 P.M.	18 15	524·2	1,430
Lyttelton „ .	5,641	2 48 „	18 23	519·8	1,400
Newcastle (Australia) . .	6,800	2 32 „	18 7	633·0	2,075
Sydney „ .	6,782	2 35 „	18 10	631·4	2,065
Kamieshi (Japan) . .	8,790	7 20 „	22 55	649·0	2,182
Hakodate „ . .	8,760	9 25 „	25 0	592·5	1,818
Kadsusa „ . .	8,939	9 50 „	25 15	604·9	1,895

[1] *Petermann's Mittheilungen*, 1877, Heft xii. S. 454, and *Nova Acta der Ksl. Leop. Carol. Deutschen Acad.* d. *Naturforscher, Band* xl. No. 9.

The mean depths represent a mean of two sets of calculations, one made with the aid of Airy's formula, and the other by Scott-Russell's formula. The result of my own investigation about this disturbance, the origin of which, by several methods of calculation, is shown to have been beneath the ocean, near 71° 5′ west long., and 21° 22′ south lat., are given on next page.

Dr. Geinitz considers that his calculated depths of the ocean and those obtained by actual soundings are in accordance, a result which is diametrically opposed to that which I have obtained.

This difference between my calculations and those of Dr. Geinitz, Hochstetter, and others, chiefly rests on the origin we have assigned for the sea waves. Dr. Geinitz, for instance, although he says that the origin of the 1877 earthquake was not on the continent but to the west in the ocean, bases all his calculations on the assumption that the *centrum* was at or near to Iquique, and the time at which that city was disturbed was the time at which the waves commenced to spread across the ocean. This time is 8.25 P.M. At this time, however, it appears that the waves must have been more than double the distance between the true origin and Iquique, from Iquique on their way towards the opposite side of the Pacific. Introducing this element into the various calculations which have been made respecting the depth of the Pacific Ocean as derived from observations on earthquake waves—which element, insomuch as the waves appear to have come in to inundate the land some time after the shock, needs to be introduced—we reduce the velocity of transit of the earthquake wave and, consequently, the resultant depths of the ocean.

In Dr. Geinitz's paper there are also some slight differences in the times at which the earthquake phe-

Origin of wave.	Longitude	Arrival of wave in Greenwich mean time (day / h. / m.)			Time taken by wave (h. / m.)		Distance from the origin in miles (calculated in great circles)	Velocity in feet per second	Depth of the ocean in feet	Height of waves	Interval between waves in minutes
Origin of wave	71 5 W.	9	12	59							
San Francisco	122 32	10	2	28	13	29	4,578	498	7,721	9 in.	
Callao	77 15	9	17	9	4	10	658	231	1,657	20 ft.	22
Iquique	70 14½	9	13	21	0	22	87	348	3,770	30 ,,	
Cobija	70 21	9	13	19	0	20	80	352	3,857	35 ,,	
Mejillones	70 35	9	13	27	0	28	108	339	3,587		15 or 45
Chanaral	71 34	9	15	26	2	27	455	272	2,309		10
Coquimbo	71 24	9	15	15	2	16	508	328	3,363		30
Valparaiso	71 38	9	16	15	3	17	695	310	3,000		
Concepcion	73 5	9	16	52	3	53	928	350	3,824		12 to 15
Honolulu	157 55	10	3	52	14	53	5,694	561	9,807	34 to 54 ft.	25
Hilo	155 3	10	3	5	14	6	5,506	563	10,217	30 or 8 ,,	3 or 15 / 18 or 27
Kahului	156 43	10	3	12	14	13	5,611	579	10,437		10
Samoa	171 41 W.	10	3	57	14	58	5,773	566	9,972		
Tauranga	176 11 E.	10	8	15	19	16	5,615	427	5,697		
Wellington	174 30	10	7	22	18	23	5,574	445	6,168	12 ft.	10
Akaroa	172 59	10	7	28	18	29	5,542	440	6,031	11 ,,	
Lyttelton	172 45	10	7	29	18	30	5,558	441	6,055		
Kameishi	140 50	10	12	37	23	38	8,844	549	9,378	6 ,,	15
Hakodate	140 50	10	14	7	25	8	8,778	512	8,169	7 ,,	20

nomena were observed at various localities. These, however, are but of minor importance. At the end of the paper by Dr. Geinitz two interesting tide gauge records are introduced, one from Sydney and the other from Newcastle. These appear to show a marked difference in the periods of the sea waves at these two places.[1]

Comparison of velocities of wave transit which have been actually observed, with velocities which ought to exist from what we know of the depth of the Pacific by actual soundings.—From a chart given in ' Petermann's Geograph. Mittheilungen,' Band xxiii. p. 164, 1877, it is possible to draw approximate sections on lines in various directions across the bed of the Pacific.

From the origin of the shock to Japan (Kameishi) the line would be as follows :—

about 7,441 miles . . . . . 15,000 feet deep
1,100 „ . . . . . 18,000 „
160 „ . . . . . 27,000 „
80 „ . . . . . 12,000 „
60 „ . . . . . 6,000 „

On account of the Tuscarora and Belkap Deeps this would be the most irregular line over which the wave had to travel.

From the origin to New Zealand (Wellington) the line would be

about 5,274 miles . . . . . 15,000 feet deep.
„ 300 „ . . . . . 12,000 „

From the origin to Samoa the line would be

about 5,773 miles . . . . . 15,000 feet deep.

From the origin to the Sandwich Islands (Honolulu) the line would be

[1] J. Milne : ' Peruvian Earthquake of May 9, 1877.' See *Trans. Seis. Soc. of Japan*, vol. ii.

almost 5,634 miles   .   .   .   . 15,000 feet deep
and   60   „   12,000   „

By Scott-Russell's rule, or, what is almost identically
the same, by Airy's general formula, we can calculate
how long it would take such waves as we have been
speaking about to travel over the different portions of
each of these lines, and by adding these times together
we obtain the time taken to travel across any one line.
I have made these calculations, but as I get in every
case answers which are too small, I think it unnecessary
to give them.

The actual times taken to travel the distances just
referred to were,

To Japan (Kameishi)   .   .   .   . 23 hr. 38 min.
„ New Zealand (Wellington) .   .   . 18 „ 23 „
„ Samoa   .   .   .   .   .   . 14 „ 58 „
„ Sandwich Islands (Honolulu)   .   . 14 „ 53 „

From San Francisco to Simoda the line is almost
3,567 miles, 3,000 fathoms deep, 840 miles, 2,500 fathoms deep, and 120 miles 1,000 fathoms deep. This
gives an average depth of about 2,854 fathoms. Bache
calculated the depth at 2,500 fathoms.

If we are to consider that, because the sea wave at
Simoda came in some time after the land shock had
been felt, the origin of this earthquake, instead of being
at Simoda, was some distance out at sea, this calculated
depth would be reduced.

CHAPTER X.

DETERMINATION OF EARTHQUAKE ORIGINS.

Approximate determination of an Origin—Earthquake-hunting in Japan—Determinations by direction of motion—Direction indicated by destruction of buildings—Direction determined by rotation—Cause of rotation—The use of time observations—Errors in such observations—Origin determined by the method of straight lines—The method of circles, the method of hyperbolas, the method of co-ordinates—Haughton's method—Difference in time between sound, earth, and water waves—Method of Seebach.

ONE of the most practical problems which can be suggested to the seismologist is the determination of the district or districts in any given country from which earthquake disturbances originate. With a map of a country before us, shaded with tints of different intensity to indicate the relative frequency of seismic disturbance in various districts, we at once see the localities where we might dwell with the least disturbance, and those we should seek if we wish to make observational seismology a study. Before erecting observatories for the systematic investigation of earthquakes in a country, it would be necessary for us, in some way or other, to examine the proposed country to find out the most suitable district. The special problem of determining *approximately* the origin or origins of a set of earthquakes would be given to us. Having made this preliminary investigation, the next point is, by means of observatories so arranged that they could always work in conjunction with each other, to determine the origins more accurately. By knowing the origin from which a set of shocks spring we know the general direction in which we may expect the most violent disturbances, and we can arrange our seismometers accordingly.

Approximate determination of origins.—In 1880 I obtained a tolerably fair idea of the distribution of seismic energy throughout Japan, by compiling the facts obtained from some hundreds of communications received from various parts of the country respecting the number of earthquakes that had been felt.

The communications were replies to letters sent to various residents in the country and to a large number of public officers. By taking these records, in conjunction with the records made by instruments, it was ascertained that in Japan alone there were certainly 1,200 shocks felt during the year, that is to say, three or four shocks per day. The greater number of these shocks were felt along the eastern coast, commencing at Tokio, in the south, and going northwards to the end of the main island. These shocks were seldom felt on the west coast. It appeared as if the central range of mountains formed a barrier to their progress. Similarly, ranges of mountains to the south-west of Tokio prevented the shocks from travelling southwards. Proceeding in this way the conclusion was arrived at that the west coast, the southern part of Japan, and the islands of Shikoku and Kiushiu, had their own local earthquakes.

Earthquake-hunting.—These preliminary enquiries having shown that the northern part of Japan was a better district for seismological observations than the southern half, the next step was to subject the northern half to a closer analysis. This analysis was commenced by sending to all the important towns, from thirty to one hundred miles distant from Tokio, bundles of postcards. These were entrusted to the local government offices with a request that each week one of these cards would be returned to Tokio stating the number of shocks felt. In this way it was quickly discovered that the majority of shakings emanated from the north and east, and seldom, if ever, passed a heavy range of mountains to the south. The barricade of postcards was then extended farther northwards, with the result of surrounding the origin of certain shocks amongst the mountains, whilst others were traced to the sea shore. By systematically pursuing earthquakes it was seen that many shocks had their origin beneath the sea—they shook all the places on the northeast coast, but it was seldom that they crossed through the mountains, forming the backbone of the island, to disturb the places on the west coast.

The actual results obtained in three months by this method of working are shown in the accompanying map, which embraces the northern half of the main island of Nipon and part of Yezo. The shaded portion of the map indicates the mountainous districts, which are traversed by ranges varying in height from about 2,000 and 7,000 feet. The dotted lines show the boundaries of the more important groups of earthquakes which were recorded.

I. is the western boundary of earthquakes, which at places to the eastward are usually felt somewhat severely. Some of these have been felt the most severely at or near Hakodadi, whilst farther south their effects have been weak. Occasionally the greatest effect has been near to Kameishi. Sometimes these earthquakes terminate along the western boundaries of III. or IV., not being able to pass the high range of mountains which separate the plain of Musashi from Kofu.

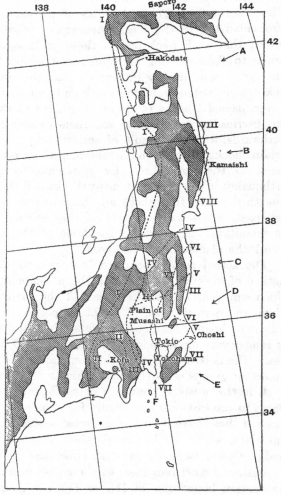

FIG. 29.—Northern Japan.   Mountainous districts shaded
with oblique lines.

II. is the boundary of a shock confined to the plain which surrounds Kofu. These earthquakes are evidently quite local. Many of the disturbances have evidently originated beneath the ocean, having come in upon the land in the direction of the arrows A or B.

III. This line indicates the boundary of a group of shocks which are often experienced in Tokio. These may come in the directions D, E, or F. It is probable that some of them originate to the eastward of Yokohama, on or near to the opposite peninsula.

IV. V. and VI. The earthquakes bounded by these lines probably originate in the directions C or D.

VII. The earthquakes bounded by this line probably come from the direction E.

VIII. This line gives us the boundary of earthquakes which may come from the direction B.

The above boundaries sometimes do not extend so far to the westward as they are shown. At other times, groups like V. and VI. extend farther to the south-west. These earthquake boundaries, which so clearly show the effects of high mountains in preventing the extension of motion, have been drawn up, not from single earthquakes, but from a large series of earthquakes which have been plotted upon blank maps, and are now bound together to form an atlas. To give an idea of the material upon which I have been working, I may state that between March 1 and March 10, 1882, I received records of no less than thirty-four distinct shocks felt in districts between Hakodate and Tokio, and for each of these it is quite possible to draw a map. In addition to the boundaries of disturbances given in the accompanying map, other boundaries might be drawn for shocks which were more local in their character. The groups which contained the greatest number of shocks are III., IV., V., VI., and VII. By

work of this description it was found that a very important group of earthquakes might be studied by a line
of stations commencing at Saporo in the north, passing
through Hakodate, down the east coast of the main island,
to Tokio or Yokohama in the south. · A further aid to
the study of this group, together with the study of an
important local group, might be effected with the help of
a few additional stations properly distributed on the plain
of Musashi, which surrounds Tokio. With this example
before us it will be recognised that the choice of sites for
a connected set of seismological observatories will often
be more or less a special problem. If earthquake stations
were to be placed in different directions around Tokio
without preliminary investigation, it is quite possible that
some of them might be so situated that they would seldom
if ever work in conjunction with the remaining observatories, and· therefore be of but little value. And this
remark must equally apply to districts in other portions
of the globe. The method is crude, and, so far as actual
earthquake origins are concerned, it only yields results
which are approximate. The crudeness and the want of
absoluteness in the results is, however, more than counterbalanced by the certainty with which we are enabled to
express ourselves with regard to such results as are obtained. Even when working with the best instruments
we have at our command, unless we are employing some
elaborate system, this method of working gives a most
valuable check upon our instrumental records, and enables
us to interpret them with greater confidence.

*Determination of earthquake origins from the direc-
tion of motion.*—If we assume that an earthquake is propagated from a centre as a series of waves, in which
normal vibrations are conspicuous, and obtain at two
localities, not in the same straight line with the origin,

and sufficiently far separated from each other, the direction of movement of these normal motions, by drawing lines parallel to these directions through our two stations, the lines would intersect at a point above the required origin. If instead of two points we had three, or, better still, a large number, the results we should obtain ought to be still more certain. Unfortunately, it seems that earthquakes seldom originate from a given point, and, further, normal motions are not always (sufficiently) prominent. Sometimes, as has already been shown in the chapters on earthquake motion, they may be non-existent. It is probable, however, that difficulties of this sort are more usually associated with non-destructive earthquakes. Mallet regards the destructive effects of an earthquake as almost solely due to normal motions. If this be true, for destructive earthquakes, the problem is shorn of many of its difficulties. In cases where normal vibrations are not prominent, where we have only transverse vibrations, motions due to the interference of normal or transverse motions, or directions of motions due to the topographical or geological nature through which the disturbance has passed, the determination of the origin of an earthquake by observations on the direction in which the ground has been moving appears to be a problem which is practically without a solution. We will, therefore, only consider the determination of the origin of those earthquakes which have predominating directions in their movements, which directions we will consider as normal ones. The question which is, then, before us, is the determination of the direction of these normal movements. First of all we may take the evidence of our senses. In exceptional cases these have given results which closely approximate to the truth, but in the majority of cases such results are not to be relied upon, as the inhabitants of a town will, for the

o

same shock, give directions corresponding to all points of
the compass. Much, no doubt, depends upon the situation of the observer, and much, perhaps, upon his temperament. If he is sitting in a room alone, and is accus-
tomed to making observations on an earthquake, on feeling
the earthquake, if he concentrates his attention on the
direction in which he is being moved, his observations
may be of value. If, however, he is not so situated, and
his attention is not thus concentrated, his opinions, unless
the motion has been very decided in its character, are
usually of but little worth.

*Direction determined from destruction of build-
ings.*—When an observer first sees a town that has been
partially shattered by an earthquake, all appears to be
confusion, and it is difficult to imagine that in such apparent chaos we are able to discover laws. If, however,
we take a general view of this destruction and compare
together similarly built buildings, it is possible to discover that similar and similarly situated structures have
suffered in a similar manner. By carefully analysing the
destruction we are enabled to infer the direction in which
the destroying forces have acted. It was chiefly by
observing the cracks in buildings, and the direction in
which bodies were overthrown or projected, that Mallet determined the origin of the Neapolitan earthquake. From
the observations given in Chapter VII. it would appear
that, with destructive earthquakes, walls which are transverse to the direction of motion are most likely to be
overturned, whilst, with small earthquakes, these walls are
the least liable to be fractured.

From a critical examination of the *general* nature of
the damage done on the buildings of a town, earthquake
observers have shown that the direction of a shock may
often be approximately determined. The direction in

which a body having a regular form like a prismatic gravestone or a cylindrical column is overturned sometimes gives the means by which we can determine the direction from which a movement came.

The rotation of bodies.—It has often been observed that almost all large earthquakes have caused objects like tombstones, obelisks, chimneys, &c., to rotate.

One of the most natural and at the same time most simple explanations is to suppose that during the shock there had been a twisting, or backward and forward screw-like motion in the ground. Amongst the Italians and the Mexicans earthquakes producing an effect like this are spoken of as 'vorticosi.' In the Calabrian earthquake, not only were bodies like obelisks twisted on their bases, but straight rows of trees seem to have been left in interrupted zigzags. These latter phenomena have been explained upon the assumption of the interference of direct waves and reflected waves, the consequence of which being that points in close proximity might be caused to move in opposite directions. Reflections such as these would be most likely to occur near to the junction of strata of different elasticity, and it may be remarked that it is often near such places that much twisting has been observed.

Another way in which it is possible for twisting to have taken place would be by the interference of the normal and transverse waves which probably always exist in an earthquake shock, or by the meeting of the parts of the normal wave itself, one having travelled in a direct line from the origin, whilst the other, travelling through variable material, has had its direction changed.

Mallet, however, has shown that the rotation may have been in many cases brought about without the supposition of any actual twisting motion of the earth—

a simple backward and forward motion being quite sufficient.   If one block of stone rests upon another, and the
two are shaken backwards and forwards in a straight line,
and if the vertical through the centre of gravity of the
upper block does not coincide with the point where there
is the greatest friction between the blocks, rotation must
take place.   If the vertical through the centre of gravity
falls on one side of the centre of friction, the rotation
would be in one direction, whilst, if on the other side, the
rotation would be in the opposite direction.

Although the above explanation is simple, and also in
many cases probably true, it hardly appears sufficient to
account for all the phenomena which have been observed.

Thus, for instance, if the stones in the Yokohama
cemetery, at the time of the earthquake of 1880, had been
twisted in consequence of the cause suggested by Mallet,
we should most certainly have found that some stones
had turned in one direction whilst others had been twisted
in another.   By a careful examination of the rotated
stones, I found that every stone—the stones being in
parallel lines—had *revolved in the same direction*, namely
in a direction opposite to that of the hands of a watch.

As it would seem highly improbable that the centre of
greatest friction in all these stones of different sizes and
shapes should have been at the same side of their centres
of gravity, an effect like this could only be explained by
the conjoint action of two successive shocks, the direction
of one being transverse to the other.

Although fully recognising the sufficiency of two transverse shocks to produce the effects which have been
observed in Yokohama, I will offer what appears to me to
be the true explanation of this phenomenon : it was first
suggested by my colleague, Mr. Gray, and appears to be
simpler than any with which I am acquainted.

If any columnar-like object, for example a prism of which the basal section is represented by A B C D (see fig. 30), receives a shock at right angles to B C, there will be a tendency for the inertia of the body to cause it to overturn on the edge B C. If the shock were at right angles to D C, the tendency would be to overturn on the edge D C. If the shock were in the direction of the diagonal C A, the tendency would be to overturn on the point C. Let us, however, now suppose the impulse to be in some direction like E G, where G is the centre of gravity of the body. For simplicity we may imagine the overturning effect to be an impulse given through G in an opposite direction—that is, in the direction G E. This force will tend to tip or make the body bear heavily on C, and at the same time to whirl round C as an axis, the direction of turn being in the direction of the hands of a watch. If, however, the direction of impulse had been E′ G, then, although the turning would still have been round C, the direction would have been *opposite* to that of the hands of a watch.

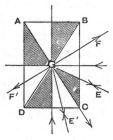

FIG. 30.

To put these statements in another form, imagine G E′ to be resolved into two components, one of them along G C and the other at right angles, G F. Here the component of the direction G C tends to make the body tip on C, whilst the other component along G F causes revolution. Similarly G E may be resolved into its two components G C and G F′, the latter being the one tending to cause revolution.

From this we see that if a body has a rectangular section, so long as it is acted upon by a shock which is parallel to its sides or to its diagonals, there ought not to be any revolution.   If we divide our section A B C D up into eight divisions by lines through these directions, we shall see that any shock the direction of which passes through any of the octants which are shaded will cause a *positive* revolution in the body—that is to say, a revolution corresponding in its direction to that of the movements of the hands of a watch; whilst if its direction passes through any of the remaining octants the revolution will be *negative*, or opposite to that of the hands of a watch.   From the direction in which any given stone has turned, we can therefore give two sets of limits between one of which the shock must have come.

Further, it will be observed that the tendency of the turning is to bring a stone, like the one we are discussing, broadside on to the shock; therefore, if a stone with a rectangular cross section has turned sufficiently, the direction of a shock will be parallel to one of its faces, but if it has not turned sufficiently it will be more nearly parallel to its faces in their new position than it was to its faces when in their original position.

If a stone receives a shock nearly parallel with its diagonal, on account of its instability it may turn either positively or negatively according as the friction on its base or some irregularity of surface bearing most influence. Similarly, if a stone receives a shock parallel to one of its faces, the twisting may be either positive or negative, but the probability is that it would only turn slightly; whereas in the former case, where the shock was nearly parallel to a diagonal, the turning would probably be great.

Determination of direction from instruments.— When speaking about earthquakes it was shown, as the result of many observations, that the same earthquake in the space of a few seconds, although it may sometimes have only one direction of motion, very often has many directions of motion. In certain cases, therefore, our records, if we assume the most permanent motions to be normal ones, give definite and valuable results. In other cases it is necessary to carefully analyse the records, comparing those taken at one station with those taken at another.

One remarkable fact which has been pointed out in reference to artificial earthquakes produced by exploding charges of gunpowder or dynamite, and also with regard to certain earthquakes, is that the greatest motion of the ground is *inwards*, towards the point from which the disturbance originated. Should this prove the rule, it gives a means of determining, not only the direction of an earthquake, but the side from which it came.

Determination of earthquake origins by time observations.—The times at which an earthquake was felt at a number of stations are among the most important observations which can be made for the determination of an earthquake origin. The methods of making time observations, and the difficulties which have to be overcome, have already been described. When determining the direction from which a shock has originated, or determining the origin of the shock by means of time observations, it has been usual to assume that the velocity of propagation of the shock has been uniform from the origin. The errors involved in this assumption appear to be as follows:—

1. We know from observations on artificial earthquakes that the velocity of propagation is greater between stations near to the origin of the shock than it is between more remote stations; and also the velocity of propagation varies with the initial force which produced the disturbance. If our points of observation are sufficiently close together as compared with their distance from the

origin of the disturbance, it is probable that errors of this description are small and will not make material differences in the general results.

2. We have reasons for believing that the transit velocity of an earthquake is dependent on the nature of the rocks through which it is propagated. Errors which arise from causes of this description will, however, be practically eliminated if our observation points are situated on an area sufficiently large, so that the distribution of the causes tending to alter the velocity of a shock balance each other. It must be remarked, that causes of this description may also produce an alteration in the direction of our shock.

Other errors which may sometimes enter into our results, when determining the origin of shocks by means of observations on velocities, are the assumptions that the disturbance has travelled along the surface from the *epicentrum* and not in a direct line from the *centrum.* Again, it is assumed that the origin is a point, whereas it may possibly be a cavity or a fissure. Lastly, if we desire extreme accuracy, we must make due allowance for the sphericity of the earth and the differences of elevation of the observing stations.

I. *The method of straight lines.*—Given a number of pairs of points A_0, A_1, B_0, B_1, C_0, C_1, &c., at each of which the shock was felt simultaneously, to determine the origin.

Theoretically if we bisect the line which joins A_0 and A_1 by a line at right angles to A_0, A_1, and similarly bisect the lines B_0, B_1, C_0, C_1, all these bisecting lines a_0, a_1, b_0, b_1, c_0, c_1, &c., ought to intersect in a point, which point will be the *epicentrum* or the point above the origin.

This method will fail, first, if A_0, A_1, B_0, B_1, C_0, C_1 form a continuous straight line, or if they form a series of parallel lines.

Hopkins gives a method based on a principle similar to the one which is here employed—namely, given that a shock arrives simultaneously at *three* points to determine, the centre. In this case, the relative positions of the three points, where the time of arrival was simultaneous, must be accurately known, and these three points must not lie in a straight line, or the method will fail. For practical application the problem must be restricted to the case of three points which do not lie nearly in the same straight line.

II. *The method of circles.*—Given the times t_0, t_1, t_2, &c., at which a shock arrived at a number of places A_0, A_1, A_2, &c., to determine the position from which the shock originated.

Suppose A_0 to be the place which the shock reached first, and that it reached A_1, A_2, A_3, &c., successively afterwards.

Let
$$t_1 - t_0 = a$$
$$t_2 - t_0 = b$$
$$t_3 - t_0 = c, \text{ &c.}$$

With A_1, A_2, A_3, &c. as centres, describe circles with radii proportional to the known qualities a, b, c, &c., and also a circle which passes through A_0 and touches these circles. The centre of the last circle will be the *epicentrum*. The radii proportional to a, b, c, &c. may be represented by the quantities ax, bx, cx, &c., where x is the velocity of propagation of the shock.

It will be observed that the times at which the shock arrived at three places might alone be sufficient. If, instead of taking the times of arrival of a shock, the arrival of a sea wave be taken, the result would be a closer approximate to the absolute truth.

It will be observed that this method is not a direct one, but is one of trial. If, however, an imaginary case be taken, and three given points of observation, A_0, A_1, A_2, be plotted on a piece of paper, it will be found that it is not a difficult matter to determine two numbers proportional to a and b which will allow you to draw two circles so that they may be touched by a third circle drawn through A_0. This problem has practically been applied in the case of the arrival of a sea wave at a number of places on the South American coast, at the time of the earthquake of May 9, 1877. This is illustrated as follows. The places which were chosen were Huanillos, Tocopilla, Cobija, Iquique, Mejillones.

In the following table the first column gives the times at which the sea wave arrived at each of these places in Iquique time; in the second column the difference between these times and the time at which it reached Huanillos is given; in the third column the distances through which a sea wave, propagated at the rate of 350 feet per second, could travel during the intervals noted in the second column is given.

	Arrival of sea wave	Time after arrival at Huanillos	Distance at 350 feet per second
	h. m.	minutes	miles
Huanillos	8 30	0	0
Tocopilla	8 32	2	8
Cobija	8 38	8	32
Iquique	8 40	10	40
Mejillones	8 46	16	64

The distances marked in the third column are used as radii of the circles drawn round the places to which they respectively refer.

The centre of the circle drawn to touch the circles of the first column, and at the same time to pass through Huanillos, is marked c.

The position from which the shock originated appears therefore to have occurred very near to a place lying in Long. 7° 15′ W. and Lat. 21° 22′ S.

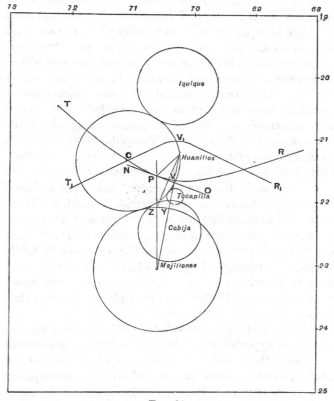

FIG. 31.

The actual operations which were gone through in making the accompanying map were as follows. First, the places with which we had to deal were represented on a map in orthographical projection, the centre of pro-

jection being near to the centre of the map. This was
done so that the measurements which were made upon
the map might be more correct than those we should
obtain from an ordinary chart where this portion of the
world was not the centre of projection. Next, a number
was taken as equal to the velocity with which the sea wave
had travelled. The first velocity taken was about 400
feet per second—this being about the velocity with which,
theoretically, it must have travelled in an ocean having
a depth equal to that indicated upon the charts—also it
seemed to have travelled at this rate from the various
times of arrival as recorded at places along the coast.
Circles were then drawn round Tocopilla, Cobija, Iquique,
and Mejillones with radii equal to 2, 8, 10, and 15, each
multiplied by (60 × 400). It was then seen *by trial* that
it was impossible to draw a single circle which should
touch four circles and also pass through Huanillos. These
four circles were, in fact, too large. Four new but smaller
circles, which are shown in the map, were next drawn, their
radii being respectively equal to the numbers 2, 8, 10,
and 16, each multiplied by (60 × 350), and it was found
that a circle, with a centre C, could be drawn which would
practically touch the four circles, and at the same time
would pass through Huanillos.

III. *The method of hyperbolas.*—The method which I
call that of hyperbolas is only another form of the method
of circles. It is, however, useful in special cases, as, for
instance, where we have the times of arrival of earthquakes
at only two stations. Between Tokio and Yokohama, at
which places I frequently obtain tolerably accurate time
records, the method has been applied on several occasions
with advantage. In the preceding example let us suppose
that the only time records which we had were for Huanillos and Mejillones, and that the wave was felt at the

latter place sixteen minutes or 960 seconds after it was experienced at the former. Calling these places H and M respectively, round M draw a circle equal to the 960 multiplied by the velocity with which the wave was propagated. It is then evident that the origin of this disturbance must be the centre of a circle which passes through H and touches the circle drawn round M. Join H M, cutting the circle round M in Y. Bisect Y H in V. It is evident that V is one possible origin for the disturbance. Next, from M, in the direction of H, draw any line M Z P; join Z H; bisect Z H at right angles by the line O P N. Because PH = PZ, it is evident that P is a second possible origin. Proceeding in this way a series of points lying to the right and left of V on the curve R V T may be found, and we may therefore say that the origin lies somewhere in the curve R V T. By increasing or decreasing our velocity we vary the position of the curve R V T, and, instead of a line on which our origin may be, we obtain a band. As the assumed velocity increases, the circle round M becomes larger, and has its limit when it passes through H, where the two arms of the curve R V T will close together and form a prolongation of the line M Y H as the assumed velocity diminishes. The circle round M becomes smaller until it coincides with the point M. At such a moment the curve R V T opens out to form a straight line bisecting M H at right angles. The curve R V T is a hyperbola with a vertex V and foci H and M. Inasmuch as PM − P H = a constant quantity. If we have the time given at which the shock or wave arrived at a third station as at Iquique, it is evident that a second hyperbola R' V' T' might be drawn with Iquique and Huanillos as foci, and that the mutual intersection of these two hyperbolas with a third hyperbola, having for its foci Iquique and Mejillones, would give the origin of the wave.

The obtaining of a mutual intersection would depend on the assumed velocity, and the accuracy of the result, like that of the method of circles, would depend upon the trials we made. The method here enunciated may be carried farther by describing hyperboloids instead of hyperbolas, the mutual intersection of which surfaces would, in the case of an earth wave, give the actual origin or *centrum* rather than the point above the origin or *epicentrum*.

IV. *The method of co-ordinates.*—Given the times at which a shock arrived at five or more places, the position of which we have marked upon a map, or chart, to determine the position on the map of the centre of the shock, its depth, and the velocity of propagation.

Commencing with the place which was last reached by the shock, call these places p, p_1, p_2, p_3, and p_4, and let the times taken to reach these places from the origin be respectively t, t_1, t_2, t_3, and t_4.

Through p draw rectangular co-ordinates, and with a scale measure the co-ordinates of p_1, p_2, p_3, and p_4, and let these respectively be a_1, b_1 ; a_2, b_2 ; a_3, b_3 ; a_4, b_4. Then if x, y, and z be the co-ordinates of the origin of the shock, d, d_1, d_2, d_3, and d_4, the respective distances of p, p_1, p_2, p_3, and p_4 from this origin, and v the velocity of the shock, we have

$$1. \quad x^2 + y^2 + z^2 = d^2 = v^2\, t^2$$
$$2. \quad (a_1 - x)^2 + (b_1 - y)^2 + z^2 = v^2\, t_1^2$$
$$3. \quad (a_2 - x)^2 + (b_2 - y)^2 + z^2 = v^2\, t_2^2$$
$$4. \quad (a_3 - x)^2 + (b_3 - y)^2 + z^2 = v^2\, t_3^2$$
$$5. \quad (a_4 - x)^2 + (b_4 - y)^2 + z^2 = v^2\, t_4^2$$

Because we know the actual times at which the waves arrived at the places p, p_1, p_2, p_3, p_4, we know the values $t - t_1$, $t - t_2$, $t - t_3$, $t - t_4$. Call these respectively m, p, q, and r. Suppose t known, then

$$t_1 = t - m$$
$$t_2 = t - p$$
$$t_3 = t - q$$
$$t_4 = t - r.$$

Subtracting equation No. 1 from each of the equations 2, 3, 4, and 5, we obtain,

$$a_1^2 + b_1^2 - 2a_1\,x - 2b_1\,y = v^2\,(t_1^2 - t^2) = v^2\,(m^2 - 2t\,m)$$
$$a_2^2 + b_2^2 - 2a_2\,x - 2b_2\,y = v^2\,(t_2^2 - t^2) = v^2\,(p^2 - 2t\,p)$$
$$a_3^2 + b_3^2 - 2a_3\,x - 2b_3\,y = v^2\,(t_3^2 - t^2) = v^2\,(q^2 - 2t\,q)$$
$$a_4^2 + b_4^2 - 2a_4\,x - 2b_4\,y = v^2\,(t_4^2 - t^2) = v^2\,(r^2 - 2t\,r)$$

Now let $v^2 = u$, and $2v^2\,t = w$.
Then

1. $\quad 2a_1\,x + 2b_1\,y + u\,m^2 - w\,m = a_1^2 + b_1^2$
2. $\quad 2a_2\,x + 2b_2\,y + u\,p^2 - w\,p = a_2^2 + b_2^2$
3. $\quad 2a_3\,x + 2b_3\,y + u\,q^2 - w\,q = a_3^2 + b_3^2$
4. $\quad 2a_4\,x + 2b_4\,y + u\,r^2 - w\,r = a_4^2 + b_4^2$

We have here four simple equations containing the four unknown quantities x, y, u, and w.

x and y determine the origin of the shock. The absolute velocity v equals $\sqrt{u}$. From v and w we obtain t. Substituting x, y, v, and t in the first equation we obtain z.

We have here assumed that the points of observation have all about the same elevation above sea level.

If it is thought necessary to take these elevations into account, a sixth equation may be introduced.

If the propagation of the wave is considered as a horizontal one, as would be done when calculating the position of the *epicentrum* or point above the origin, by means of the times of arrival of a sea wave, the ordinate z of the first five equations would be omitted. Working in this way the resulting four equations, viz.

$$2a_1\,x + 2b_1\,y + um^2 - wm^2 = a_1^2 + b_1^2,$$
$$\text{\&c.} \qquad \text{\&c.} \qquad \text{\&c.}$$

remained unchanged.

Applying this method to the same example as that

used as illustration for the two previous methods, we
obtain for the co-ordinates of Mejillones, Iquique, Cobija,
Tocapilla, and Huanillos, measured in geographical miles,
and the times in Iquique time at which the wave reached
each, as given in the following table; *ox* and *oy* being,
drawn through Mejillones.

		Co-ordinates		Time of arrival
		OX	OY	h. m.
Mejillones . .	*a* or 0	*b* or 0	8 46 p. m.
Iquique . . .	a_1 or 150	b_1 or 96	8 40 „
Cobija . . .	a_2 or 36	b_2 or 14	8 38 „
Tocopilla . .	a_3 or 66	b_3 or 31	8 32 „
Huanillos . .	a_4 or 102	b_4 or 58	8 30 „

From this data we find the co-ordinates x and y of
this origin to be 85·8 and 56·7; whilst the velocity of
propagation = 45 feet per second.

Measuring these ordinates upon the map, we obtain a
centre lying very near Long. 71° 5′ W. and Lat. 21° 22′
S., a position which is very near to that which has already
been obtained by other methods.

If instead of Huanillos we substitute the ordinates
and time of arrival of the sea wave for Pabalon de Pica,
another point for the origin will be obtained lying farther
out at sea.   To obtain the best result, the method to be
taken will evidently be, first to reject those places at
which it seems likely that some mistake has been made
with the time observations, and then with the remaining
places to form as many equations as possible, and from
these to obtain a mean value.   This is a long and tedious
process, and as the time observations of this particular
earthquake are probably one and all more or less inaccurate, it is hardly worth while to follow the investigation
farther.

In this example, as in the preceding ones, it will be

observed that it has been sea waves that have been dealt with, rather than earth vibrations. It is evident, however, that these latter vibrations may be dealt with in a similar manner.

In these determinations it will also have been observed that it is assumed that the disturbance has radiated from a centre, and, therefore, approached the various stations in different directions. If we assume that we have three stations very near to each other as compared with their distances from the origin, so that we can assume that the wave fronts at the various stations were parallel, the determination of the direction in which the wave advanced appears to be much simplified. The determination of the direction in which a wave has passed across three stations was first given by Professor Haughton.

Haughton's method.—Given, the time of an earthquake shock at three places, to determine its horizontal velocity and coseismal line.

The solution of this is contained in the formula

$$\tan \phi = \frac{a \, (t_2 - t_1) \sin \beta}{c \, (t_3 - t_2) + a \, (t_2 - t_1) \cos \beta}$$

When A, B, and C are three stations at which a shock is observed at the times t_1, t_2, and t_3; a, b, and c are the distances between A, B, and C, and ϕ is the angle made by the coseismal lines $x \, \text{A} \, x$, $y \, \text{B} \, y$, and the line A B, which are assumed to be parallel.

This I applied in the case of the Iquique earthquake, but owing to the smallness of the angles between the three stations A, B, and C, the result was unsatisfactory. The problem ought to be restricted, first, to places which are a long distance away from a centre, and, secondly, to places which are not nearly in a straight line. This problem may be solved more readily by geometrical methods.

P

Plot the three stations A, B, and C on a map, join the two
stations between which there was the greatest difference
in the time observation. Let these, for example, be A
and C. Divide the line A C at point D, so that A D : D C as
the interval between the shock felt at A and B is to the
interval between the shock felt at B and C. The line B D
will be parallel to the direction in which the wave advanced.

*The difference in time of the arrival of two disturb-
ances.*—In the various calculations which have been made
to determine an origin based on the assumption of a
known or of a constant velocity, we have only dealt with
a single wave, which may have been a disturbance in the
earth or in the water. A factor which has not yet been
employed in this investigation is the difference in time
between the arrival of two disturbances; one propagated,
for instance, through the earth, and the other, for
example, through the ocean. The difference in the times
of the arrival of two waves of this description is a
quantity which is so often recorded that it is well not to
pass it by unnoticed. To the waves mentioned we might
also add sound waves, which so frequently accompany
destructive earthquakes, and, in some localities, as, for
instance, in Kameishi, in North Japan, are also commonly
associated with earthquakes of but small intensity. It
was by observing the difference in time between the
shaking and the sound in different localities that Signor
Abella was enabled to come to definite conclusions regarding the origin of the disturbances which affected the
province of Neuva Viscoya in the Philippines, in 1881;
the places where the interval of time was short, or the
places where the two phenomena were almost simultaneous, being, in all probability, nearer to the origin
than when the intervals were comparatively large. I
myself applied the method with considerable success

when seeking for the origin of the Iquique earthquake of 1877. The assumptions made in that particular instance were, first, that the velocity of the disturbance through the earth was known, and, secondly, that the velocity with which a sea wave was propagated was also known.

A method similar to the above was first suggested by Hopkins. It depended on the differences of velocity with which normal and transversal waves are propagated.[1]

Seebach's method. —To determine the true velocity of an earthquake, the time of the first shock, and the depth of the centre.

Let the straight line M, m_1, m_2, m_3 represent the surface of the earth shaken by an earthquake. For small earthquakes, to consider the surface of the earth as a plane will not lead to serious errors. If an earthquake originates at C, then to reach the surface at M it traverses a distance h in the time t. To reach the surface at M_1, it traverses a distance $h + x_1$ in a time t_2. If v equals the velocity of propagation,

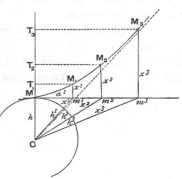

FIG. 32.

$$\text{then } t = \frac{h}{v}, \quad t_1 = \frac{h + x_1}{v},$$

$$t_2 = \frac{h + x_2}{v}, \text{ &c.}$$

Seebach now says that *if we have given the position of* M *or epicentrum of the shock,* and draw through it rectangular axes like M m_3 and M T_3, and lay down on

[1] *Report of British Association*, 1847, p. 84.

M m_3 in miles the distances from M of the various stations which have been shaken, and in equal divisions for minutes lay down on M T_3 the differences of time at which M, m_1, m_2, &c. were shaken, then $M_1 T_1$, $M_2 T_2$, &c. are the co-ordinates of points on an hyperbola. The degree of exactness with which this hyperbola is in any given case constructed is a check upon the accuracy of the time observations and the position of the *epicentrum*. The apex of the hyperbola is the *epicentrum*.

The intersection of the asymptote with the ordinate axis is the time point of the first shock, which, because the scale for time and for space were taken as equal, gives the absolute position of the *centrum*. This intersection is shown by dotted lines. Knowing the position of the *centrum*, we can directly read from our diagram how far the disturbance has been propagated in a given time.

CHAPTER XI.

THE DEPTH OF AN EARTHQUAKE CENTRUM.

The depth of an earthquake centrum—Greatest possible depth of a
earthquake—Form of the focal cavity.

Depth of centrum.—The first calculations of the depth at
which an earthquake originated were those made by Mallet
for the Neapolitan earthquake of 1857. These were made
on the assumption that the earth wave radiated in straight
lines from the origin, and, therefore, at points at different
distances from the *epicentrum* it had different angles of
emergence. These angles of emergence were chiefly calculated from the inclination of fissures produced in certain
buildings, which were assumed to be at right angles to the
direction of the normal motion. If we have determined
the *epicentrum* of an earthquake and the muzoseismal
circle, and make either the assumption that the angle of
emergence in this circle has been 45° or 54° 44' 9" (see
page 54), it is evidently an easy matter by geometrical
construction to determine the depth of the *centrum*.
Hofer followed this method when investigating the earthquake of Belluno.

Other methods of calculation which have been employed
are based on time observations, as, for instance, the method
of Seebach, the method of co-ordinates, the method of hyperboloids or spheres (see pages 200–212).

By means of a number of lines parallel to twenty-six

angles of emergence, drawn in towards the seismic vertical, Mallet found that twenty-three of these intersected at a depth of $7\frac{1}{8}$ geographical miles. The maximum depth was $8\frac{1}{8}$ geographical miles, and the minimum depth $2\frac{3}{4}$ geographical miles.

The mean depth was taken at a depth of $5\frac{3}{4}$ geographical miles where, within a range of 12,000 feet, eighteen of the wave paths intersected the seismic vertical.

The point where these wave paths start thickest is at a depth not greater than three geographical miles, and this is considered to be the vertical depth of the focal cavity itself.

For the Yokohama earthquake of 1880, from the indications of seismometers, and by other means, certain angles of emergence were obtained, leading to the conclusion that the depth of origin of that earthquake might be between $1\frac{1}{2}$ and 5 miles.

Possibly, perhaps, the earthquake may have originated from a fissure the vertical dimensions of which was comprised between these depths.

A source of error in a calculation of this description is that the vertical motions may have been a component of transverse motions or perhaps due to the slope of surface waves.

The following table of the depths at which certain earthquakes have originated has been compiled from the writings of several observers.

		In feet		
		Minimum	Mean	Maximum
Rhineland . .	1846 (Schmidst)		127,309	
Sillien . . .	1858 (Schmidst)		86,173	
Middle Germany .	1872 (Seebach)	47,225	58,912	70,841
Herzogenrath .	1873 (Lasaulx)	16,553	36,516	56,477
Neapolitan . .	1857 (Mallet)	16,705	34,930	49,359
Yokohama . .	1880 (Milne)	7,920	17,260	26,400

A table similar to this has been compiled by Lasaulx.[1]

With the exception of the determination for the two last disturbances these calculations have been made with the assistance of the method of Seebach, which depends, amongst other things, on the assumptions of exact time determinations, direct transmission of waves from the *centrum*, and a constant velocity of propagation.

Admitting that our observations of time are practically accurate, it appears that the other assumptions may often lead to errors of such magnitude that our results may be of but little value.

From what has been said respecting the velocity with which earth disturbances are propagated, it seems that these velocities may vary between large limits, being greatest nearest to the origin.

If we refer to Seebach's method, we shall see that a condition of this kind would tend to make the differences in time between various places, as we recede from the *epicentrum*, greater than that required for the construction of the hyperbola. The curve which is obtained would, in consequence, have branches steeper than that of the hyperbola, and the resultant depth, obtained by the intersection of the asymptotes of this curve with the seismic vertical, indicates an origin which may be much too great.

Another point worthy of attention, which is common to the method of Mallet as well as to that of Seebach, is the question whether the shock radiates directly from the origin, or is propagated from the origin more or less vertically to the surface, and then spreads horizontally. We know that earthquakes, both natural and artificial, may be propagated as undulations on the surface of the ground, and that the vertical motion of the latter, as

[1] *Das Erdbeben von Herzogenrath, &c.*, p. 134.

testified by the records of well-constructed instruments, has no practical connection with the depth from which the disturbance originated.

In cases like these, the direction of cracks in buildings, and other phenomena usually accredited to a normal radiation, may in reality be due to changes in inclination of the surface on which the disturbed objects rested. When our points of observation are at a distance from the *epicentrum* of the disturbance which, as compared with the depth of the same, is not great, calculations or observations based on the assumption of a direct radiation of the disturbance may possibly lead to results which are tolerably correct. The calculations of Mallet for the Neapolitan earthquake appear to have been made under such conditions.

For smaller earthquakes, and for places at a distance from the seismic vertical of a destructive earthquake, the results which are deduced from the observations on shattered buildings, and all observations based upon the assumption of direct radiation, we must accept with caution.

Another error which may enter into calculations of this description is one which has been discussed by Mallet at some length. This is the effect which the form and the position of the focal cavity may have upon the transmission of waves.

Should the impulse originate from a point or spherical cavity, then we might, in a homogeneous medium perhaps, regard the isoseismals as concentric circles, and expect to find that equal effects had been produced at equal distances from the *epicentrum*. Should, however, this cavity be a fissure, it is evident that even in a homogeneous medium the inclination of the plane of such a cavity will have considerable effect upon the form of the waves which would radiate from its two walls.

For example, let it be assumed that the first impulse of an earthquake is due to the sudden formation of a fissure, rent open from its centre, and that the waves leave the walls at all points normal to its surface. Then, as Mallet points out, it is evident that the disturbance will spread out in ellipsoidal waves, the greatest axis of which will be perpendicular to the plane of the fissure.

By taking a number of cases of fissures lying in various directions and drawing the ellipsoidal waves which would result from an elastic pressure, like that of steam suddenly admitted into such cavities, the differences in effect which would be simultaneously produced by these waves on reaching the surface can be readily understood. The following example of an investigation on this subject will serve as an example to illustrate the general nature of the many other cases which might be taken.

Let a disturbance simultaneously originate from all points of the fissure *f f*. This will spread outwards in ellipsoidal shells to the surface of the earth *e e*. The major axis of these ellipsoidal shells will be the direction of greatest effect. In the direction *c d* the waves will plunge into the earth, and places to the right side of the fissure

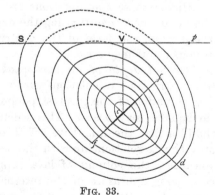

FIG. 33.

will, to use an expression due to Stokes, when speaking of analogous phenomena connected with sound, be in *earthquake shadow*. The same expression has been employed, somewhat differently, when speaking of the effects produced on buildings.

For places, like *s* and *p*, situated at equal distances from the seismic vertical, it is evident that the intensity of the shock will be different, and also its time of arrival. It will also be observed that the isoseismals will take the form of ovals or distorted ellipses, the larger or fuller end of which being to the left of the fissure.

Other cases, like those just given, which are discussed by Mallet in his account of the Neapolitan earthquake, are where the fissure forms the division between materials of different elasticities. In the hard and more elastic material the waves will be more crowded, the velocity of a wave particle will be greater, and the transit will be quicker than in the less elastic medium.

The result is that the distance of equal effect from the seismic vertical will be greatest in the direction of the more compressible material.

Unless these considerations are kept carefully before the mind when investigating the depth and, we may add, the position and form of the centrum of an earthquake, serious errors may arise.

Greatest depth of an earthquake origin.—A curious but instructive calculation which Mallet made was a determination of the greatest possible depth at which an earthquake may occur. This calculation is based upon the idea that the impulsive effect of an earthquake has an intimate relationship with the height of neighbouring volcanoes, the column of lava supported on a volcanic cone being a measure of the internal pressure tending to rupture the adjacent crust of the earth.

Michell, in 1700, virtually propounded this idea, when he suggested that the velocity of propagation of an earthquake was related to the height of such a column.[1]

[1] *Phil. Trans.* vol. li.

Mallet showed that there was probably considerable truth in such a supposition by appealing to the results of actual observation. The pressure gauge of the Neapolitan district would be Vesuvius, the height of which has in round numbers varied between 3,500 to 4,000 feet. One of the most destructive earthquakes in this district— namely, the one of 1857—projected bodies with an initial velocity of about fifteen feet per second. The Riobamba earthquake, which projected bodies with an initial velocity of eighty feet per second, appears to have been the most violent earthquake, so far as its impulsive effort is concerned, of which we have any record. It occurred amongst the Andes, where there are volcanoes from 16,000 to 21,000 feet in height.

Comparing these two earthquakes together, we see that the Riobamba shock had a destructive power 5·33 times that of the Neapolitan shock, and we also see that the Riobamba volcanoes were about 5·33 times higher than Vesuvius. The accordance in these quantities is certainly interesting, and tends to substantiate the idea that volcanoes are barometrical-like pressure gauges of a district.

Carrying the argument still further, Mallet says that if the depth of origin of earthquakes were the same, then the *area of disturbance* would, for like formations and configuration of surface, be a measure of the earthquake effort, and also some function of the velocity of the wave. From this we may generally infer ' that earthquakes, like that of Lisbon, which have a *very great area* of sensible disturbance, have also a very deep seismal focus, and also the greatest depth of seismal focus within our planet is probably not greater than that ascertained for this Neapolitan earthquake, multiplied by the ratio that the velocity of the Riobamba wave bears to that of its wave,

or, what is the same thing, by the ratio of the altitudes
of the volcanoes of the Andes to that of Vesuvius.'

Now, as the depth of the Neapolitan shock may be
taken at 34,930 feet, the greatest probable depth of origin
of any earthquake impulse occurring in our planet is
limited to 5·333 × 4,930 feet, or 30·64 geographical miles.

Ingenious as this argument is, we can hardly admit it
without certain qualifications.

First, we are called upon to admit the identity of the
originating cause of the volcano and the earthquake—as
to what may be the originating cause of earthquakes we
have yet to refer, but certainly in the case of particular
earthquakes, as, for instance, those which occur in countries
like Scotland, Scandinavia, and portions of Siberia, the
direct connection between these phenomena are not at first
sight very apparent.

Secondly, even if we admit the identity of the origin
of these phenomena, it is not difficult to imagine that
the fluid pressure brought to bear upon certain portions
of the crust of the earth may possibly in many instances
be infinitely greater than that indicated by the height
of the column of liquid lava in the throat of a volcano,
the true height of which we are unable to obtain.
Further, in certain instances such a column only appears
to be a measure of the pressure upon the crust of the
earth in the immediate vicinity of the cone.

Thus, in the Sandwich Islands, we have lava standing
in the throat of the volcano of Mouna Loa 10,000 feet
higher than it stands in the crater Kilauea, only twenty
miles distant. That these columns should be measures
of the same pressure, originating in a general subterranean
liquid layer with which they are connected, is a supposition
difficult to satisfactorily substantiate.

Another measure of the impulsive efforts which sub-

terranean forces may exert upon the crust above them is
evidently the height to which volcanoes eject materials.
Cotopaxi is said to have hurled a 200-ton block of stone
nine miles.   Sir W. Hamilton tells us that in 1779 Vesuvius shot up a column of ashes 10,000 feet in height;
and Judd tells us that this same mountain in 1872 threw
up vapours and rock fragments to the enormous height
of 20,000 feet.   This would indicate an initial velocity of
1,131 feet per second.

Notwithstanding Mallet's calculation that thirty miles
is the limiting depth for the origin of an earthquake, the
origin of the Owen's Valley earthquake of March 1872
was estimated as being at least fifty miles.[1]

Form of the focal cavity. — Among the various
problems which are put before those who study the
physics of the interior of our earth it would at first sight
appear that there was none more difficult than the attempt
to determine the form of the cavity, if it be a cavity, from
which an earthquake originates.   Almost all investigators
of seismology have recognised that the birthplace of an
earthquake is not a point, and have made suggestions
about its general nature.   The ordinary supposition is
that the earthquake originates from a fissure, and if the
focus of a disturbance could be laid bare to us it would
have the appearance of a fault such as we so often see
exposed on the faces of cliffs.

A strong argument, tending to demonstrate that some
of the shakings which are felt in Japan are due to the
production of such fissures, is the fact that the vibrations
which are recorded are transverse to a line joining the
point of observation and the district from which, by time
observations, we know the shock to have originated.   The
most probable explanation of this phenomena appears to

[1] See *Am. Jour. Sci.* 1872.

be that one mass of rock has been sliding across another
mass, giving rise to shearing strains, and producing waves
of distortion.

The first seismologist who attacked the problem of
finding out the dimensions and position of such a fissure
was Mallet, when working on the Neapolitan earthquake
of 1857. The reasons that the origin should, in the first
place, have been a fissure, rather than any other form of
cavity, was that such a supposition seemed to be *a priori*
the most probable, and, further, that it afforded a better
explanation of the various phenomena which were observed,
than that obtained from any other assumption.

The method on which Mallet worked to determine
the form and position of the assumed fissure, which
method was subsequently more or less closely followed
by other investigators, was as follows :—

From an observation of the various phenomena produced upon the surface of the disturbed area, a map of
isoseismals was constructed. These were seen, as has
been the case with many earthquakes, not to distribute
themselves in circles round the *epicentrum*, but as distorted
oval or elliptical figures, the major axes of which roughly
coincided with each other. Further, the *epicentrum*, did
not lie in the centre of these ovals, but was near to the
narrow end where they converged.

This at once showed, if the reasoning respecting the
manner in which waves are propagated from an inclined
fissure be correct, that the fissure was at right angles to
the major axis of the curves, dipping from their narrow
end downwards, in the direction of their larger widespread ends.

The next weapon which Mallet employed to attack
this problem was the sound which was heard at different
points round about the focus. These sounds appear to have

been of the nature of sudden explosive reports accompanied by rushing, rolling sounds. The form of the area in which these sounds were heard was closely similar to that of the first two isoseismals. Except in the central area of great disturbance, no sound was heard to accompany the shock.

Those at the northern and southern extremity of the sound area all described what they heard as a 'low, grating, heavy, sighing rush, of twenty to sixty seconds' duration.' Those in the middle and towards the east and west boundaries of this area described a sound of the same tone, but shorter and more abrupt, and accompanied with more rumbling.

The nature of the arguments which were followed in discussing the sound observations will be found in the chapter relating to these phenomena.

A portion of the argument which it is difficult to follow relates to the maximum rate at which it can be supposed possible for a fissure to be rent in rocks, which rate depends on the density and elasticity of these rocks and other constant factors.

Next it was observed that the paths of the waves drawn on the surface, although generally intersecting in a point, did not do so absolutely, but along a line passing through the main focus some $7\frac{1}{2}$ miles in length. This, coupled with the observations of sounds, led to the supposition that the centre of disturbance, considered horizontally, originated along a curved line passing through the chief focus and the various intersections of the wave paths.

The last phenomena brought forward to assist in the solution of this interesting problem were a study of the tremulous movements that preceded and followed the shock, and their relation to the sound phenomena.

If the earthquake originated by the formation of a fissure, after the rending has gone on for a certain time the focal cavity is enlarged to a certain extent, and the great shock takes place. This would be followed by concluding tremulous waves. A succession of phenomena like those accompanied the shock about which Mallet writes.

By observations such as these, coupled with what has been said about the maximum and mean depths of the focal cavity, Mallet came to the conclusion that the focal cavity was a fissure, the rending open of which had produced the earthquake. The vertical dimensions of this cavity were not more than 5·3 miles, but were probably limited to three miles.

From the intersection of the wave paths upon the surface and the observed emergences, this fissure followed horizontally a curve of double flexure, about nine geographical miles in length. The area of this fissure was twenty-seven geographical miles. The time of rending it open in Apennine limestone would be about $7\frac{1}{2}$ seconds, which should be the same as the period during which tremors were felt. The time actually recorded was six or eight seconds.

Briefly, this is, then, the line of reasoning which was followed by Mallet in an investigation the results of which are as interesting as they are startling. Since the line of investigation has been opened, and the existence of new problems has been indicated, other investigators, although not exactly following Mallet's method in all their details, have, when endeavouring to attain the same ends, employed similar weapons.

Thus, for example, Seebach, when determining the depth and nature of the origin of the earthquake of Middle Germany, reasoned somewhat as follows:—

Had the origin been more or less of a spherical cavity, then the region of most violent disturbance upon the surface would, according to a theorem we have already mentioned, have been upon or near a circle of about 8·8 miles in radius round the *epicentrum*. This region, however, was found by observation to lie along a curved band about forty miles in length, altogether on one side of the *epicentrum*.

To explain this anomaly Seebach followed Mallet, and assumed that the origin was not a spherical cavity, but a fissure.

The depth and strike of this fissure was determined by the observation that the area of greatest disturbance was along a curved line lying radial to the *epicentrum*. Such a condition it was assumed indicated that the fissure of origin must be inclined towards this area of greatest disturbance. A line was then drawn from this area to the *centrum*. A second line at right angles to this one gave the dip of the fissure.

Höfer, when working on the earthquake of Belluno, came to the conclusion that the disturbance originated from two faults meeting each other at an angle of 60°. In this determination he was chiefly influenced by the form of a certain homoseist which was of the form of an elongated ellipse met on one side by a second ellipse, the principal axes of the two ellipses giving the strike of the two faults.

CHAPTER XII.

DISTRIBUTION OF EARTHQUAKES IN SPACE AND TIME.

General distribution of earthquakes—Occurrence along lines—Examples of distribution—Italian earthquake of 1873—In Tokio—
Extension of earthquake boundaries—Seismic energy in relation to
geological time ; to historical time—Relative frequency of earthquakes—Synchronism of earthquakes—Secondary earthquakes. ,

General distribution of earthquakes.—The records of
earthquakes collected by various seismologists lead us
to the conclusion that at some time or other every
country and every ocean in the world has experienced
seismic disturbances. In some countries earthquakes are
felt daily, and from what will be said in the chapter on
earth pulsations it is not unlikely that every large earthquake might with proper instrumental appliances be
recorded at any point on the land surfaces of our globe.
The area over which any given earthquake extends is
indeterminate. The area over which an earthquake is
sensible is sometimes very great. The Lisbon shock of
1755 is estimated as having been sensible over an area of
3,300 miles long and 2,700 miles wide, but in the form
of tremors and pulsations it may have shaken the whole
globe.

The regions in which earthquakes are frequent are
indicated in the accompanying map, which, to a great
extent, is a reproduction of a map drawn by Mallet. The
regions coloured with the darkest tint are those where

great earthquakes are the most frequent. The actual number of earthquakes which have been felt in the differently coloured areas are given, when speaking of the relation of seismic energy to season.

When looking at this chart it must be remembered that if we were to make a detailed map of any one of the different countries where earthquakes are frequent, we should find in it all the differences that we observe in the general chart. For instance, one portion of Japan, where perhaps sixty shocks are felt per year, would be coloured with a dark tint, whilst other portions of the same country, where there is only one slight shaking felt every few years, would be left almost uncoloured. The black dots indicating the position of volcanic vents are even more general in their signification than the tinted areas. Professor Haughton gives for the world a list of 407 volcanoes, 225 of which are active. These numbers are the same as those given by A. von Humboldt. Of the active volcanoes 172 are on the margin of the Pacific, and of the total number eight are in Japan. From my own observations in Japan independently of the Kurile Islands, I have enumerated fifty-three volcanoes which are either active or have been active within a recent period. In a few years' time this list will probably be increased. I mention this fact to show how very imperfect our knowledge is respecting the number of volcanic vents existing on our globe. If we were in a position to indicate the volcanoes which had been in eruption during the last 4,000 years, the probability is that they would number several thousands rather than four or five hundred.

An inspection of the map shows that earthquakes chiefly occur in volcanic and mountainous regions. The most earthquake-shaken regions fo the world form the

boundaries of the Pacific ocean. It may be remarked
that these boundaries slope beneath the neighbouring
ocean at a much steeper angle than the boundaries of
countries where earthquakes occur but seldom. The
coasts of South America, Kamschatka, the Kuriles, Japan,
and the Sandwich Islands, for example, have slopes beneath
the Pacific from one in twenty to one in thirty. The
coasts of Australia, Scandinavia, and the eastern parts of
South America, where earthquakes are practically unknown, have slopes from one in fifty to one in two
hundred and fifty. Many earthquakes have taken place
in mid-ocean. In the Atlantic Ocean M. Perrey has
given about 140 instances of such occurrences.

The majority of the earthquakes which shake Japan
appear to have their origin in the neighbouring ocean.
If we could draw a map of earthquake origins, it is probable that the greater number of the marks indicating
these origins would be found to be suboceanic and along
lines parallel to the shores of continents and islands
which rise steeply from the bed of deep oceans. In
countries like Switzerland and India, our marks would
hold a relationship to the mountains of these countries.[1]
Looking at the broad features of the globe, we see on its
surface many vast depressions. Some of these saucer-like
hollows form land surfaces, as in central Asia. The
majority of these, however, are occupied by the oceans.
Active volcanoes chiefly occur near the rim of the hollows
which have the steepest slopes. The majority of earthquakes probably have their origin on or near the bottom
of these slopes. To these, however, there are exceptions,
as for instance the earthquakes in the Alps, in the hills of

---

[1] David Milne says that ' out of 110 shocks recorded in England,
thirty-one originated in Wales, thirty-one along the south coast of
England, fourteen on the borders of Yorkshire and Derbyshire, and five
or six in Cumberland.'

Scotland, and the shakings which are occasionally felt in countries like Egypt. The earthquakes which shake the borders of the Pacific have their origins in, and their effects are almost exclusively felt on, the sides of the bounding ridge facing this ocean. In Japan it is the eastern sides of the islands which suffer, the western side being almost as free from these convulsions as England.

Similar remarks may be made about the eastern side of South America, especially the southern portion of the continent. At Buenos Ayres, for example, there has been no disturbance since Mendoza was destroyed, some twenty years ago. In British Guiana slight shocks are occasionally felt in the low delta which forms the settled portion of the colony, but they are extremely rare.

Disturbances in lines or zones.—It has often been observed that disturbances are propagated along the length of mountains or valleys, and it is but seldom that earthquakes cross them transversely. Thus the valleys of the Rhone, the Rhine, and the Danube are lines along which disturbances travel.

The major axes of the elliptical areas of disturbances which have shaken India have a general direction parallel to the valley of the Ganges along the flanks of the Himalayas.

The disturbances which have shaken London appear to have been chiefly east and west, or along the valley of the Thames. In South America the line of disturbance is along the western sides of the Andes. Another line is along the northern coast of the continent through Andalusia and Caraccas towards the Antilles and Trinidad. The shocks of the Pyrenees are chiefly felt along the southern side of these mountains. In the middle and on the northern side they are but seldom felt. This propagation in lines or zones may in certain cases be apparent rather

than real. Thus the north and south ranges of mountains
in Japan are mountains almost simultaneously shaken
along their eastern flanks, giving the impression that an
earthquake had originated simultaneously from a fissure
parallel to this line, or else, starting at one end, had run
down their lengths. Time observations have, however,
shown that such disturbances had their origin at some
distance in the ocean, and, travelling inwards, had reached
all points on the flanks of these mountains almost simultaneously. The same explanation will probably hold for the
so-called linear disturbances of western South America.

All earthquake disturbances have probably a tendency
to radiate from their source, and are only prevented from
doing so by meeting with heavy mountainous districts,
which by their mass and structure absorb the energy communicated to them. Much energy is also lost by emergence on the open flanks of a range of mountains. Rather
than say that high mountains often bound the extension
of an earthquake, or that earthquakes appear to run
along the flanks of such mountains, we might say that
earthquakes have boundaries parallel to the strike of the
rocks in a given district, that such a direction is the one
in which the propagation is the easier.

Rossi is of opinion that volcanic fractures play an
important part in governing the distribution of seismic
disturbances. When a volcano is formed, a series of
starlike fractures are formed round the central crater.
Secondary craters may indicate the line of these fissures.
The mountains about Rome are regarded as typical of this
radial structure. The more distant the secondary craters
are from the centre of the system, the smaller will they
be, and also the younger. If two fissures intersect we
get a larger crater at the junction. Earthquakes are
propagated along the direction of these fissures, whilst the

rising and falling of these lips throw off transverse waves. Rossi adduces observations which appear to meet with explanation on such suppositions.

Suess, who has written upon the earthquakes of lower Austria, shows how the majority of the disturbances have had their origin along certain lines which form a break in the continuity of the Alps. One line runs north-east from Bruck towards Vienna. Near Wiener Neustadt, where the greatest number and heaviest shocks have occurred, this line is met by a north-north-west line crossing the Danube and following the valley of the river Kamp.[1] Hoeffer has drawn similar lines from the head of the Adriatic, one set running north-north-east to intersect near Litschau, and the other north-north-west to intersect near Frankfort in the valley of the Rhine.[2]

Examples of distribution.—A curious example of the distribution of seismic movement is that of the earthquake of March 12, 1873, worked out by Professor P. A. Serpieri. This earthquake appears to have been simultaneously felt on the Dalmatian coast and in central Italy, in a region lying north-east from Rome and south-east from Florence. In both of these areas the motion was from south-east to north-west. The ,shock then radiated from the central Italian regions, so that at places on the western shore of the Adriatic it was felt after it had been felt on the Dalmatian coast.

Many explanations might be offered for this peculiar distribution of seismic activity. Possibly the shock originated at a great depth beneath the bed of the southern part of the Adriatic, and by following existing lines of weakness simultaneously reached the surface of the earth in central Italy and Dalmatia.

[1] E. Suess, *Die Erdbeben Niederösterrreiches.*
[2] H. Hoeffer, *Die Erdbeben Kärnten's.*

In Tokio, which is built partly on a flat plain, partly
in valleys denuded from a low tableland, and partly on
the spurs of the tableland itself, the distribution of
earthquakes is a subject yet requiring attention. Sometimes it has happened that persons in one house have
been sufficiently alarmed to escape into the open air,

Earthquake of March, 12th 1873.
(P.A.Serpieri)

FIG. 34.

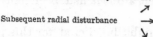

Areas almost simultaneously struck from S.E. to N.W.

Subsequent radial disturbance

whilst others, not more than a mile distant, have not been
aware that the city had been shaken.

Extension of earthquake boundaries.—Natural ob-
structions which may be sufficient to retard small earthquakes may in certain instances not be found · sufficient
to retard the larger disturbances. Thus the shocks of

Calabria are usually only felt on the western side of the Apennines, but instances have occurred when they have crossed this barrier. In 1801 the earthquake of Cumana crossed a branch of the coast range.

Sometimes earthquake boundaries give way, and countries which they sheltered subsequently become exposed to all disturbances. The true explanation of this is probably in a shifting of the centre of seismic activity. Thus up to December 14, 1797, although Cumana was often devastated, the peninsula of Araya was not hurt. On this date Araya commenced to suffer, and has continued to suffer ever since.

Fuchs gives an example of the movement of a seismic centre in the case of the Calabrian earthquake. The first shock commenced near Oppiedo, the second shock commenced four or five miles farther to the north, and the third shock had its origin five or six miles still farther, near to Girifalco.

CHAPTER XIII.

DISTRIBUTION OF EARTHQUAKES IN TIME (*continued*).

Seismic energy in relation to geological time. —If we admit that seismic energy is only a form of volcanic energy, it must also be admitted that any cause tending to produce a general decrease in the amount of the latter will also produce an alteration in the amount of the former.

The nebular hypothesis of Laplace tells us that the solar system is the result of the whirling of a heated gaseous mass, which as it cooled continually contracted and consequently whirls the faster. With this hypothesis before us, we understand why all the planets and their satellites have a similarity in the directions of their movements, why they revolve nearly in the same plane, in orbits nearly circular, why some have a flattened figure and are surrounded by rings or belts, why the exterior planets should have a greater velocity of rotation, a greater number of satellites, and a less density as compared with the interior planets, the similarity of the elements in meteoric stones, the sun, the stars, and those found upon our earth, and lastly why there should be an increase in temperature as we descend into our earth.[1] This increase in temperature as we descend into the earth

---

[1] *Six Lectures on Physical Geography*, by Rev. S. Haughton, F.R.S., chap. i.

as deduced from many observations appears to be about
1° F. for every fifty or sixty feet of descent.

To explain this and other kindred phenomena it is
assumed that the earth was once very much hotter than
it is at present, and to reach its present stage it has been
gradually cooling. As the laws of cooling are perfectly
known, to calculate how many years it must have taken a
body like our earth to cool down to its present temperature is a definite problem. Sir William Thomson, starting with the temperature of 7,000° F., when all the rocks
of the earth must have been molten and a skin or crust
upon the surface, such as is so quickly produced upon the
surface of molten lava, finds by calculation that the time
taken to reach the present temperature must have been
about one hundred million years. Into this period he
and other physicists desire to compress the history of all
the stratified deposits. Geologists find this period too
short. Others seeking to reconcile the views of physicists
and geologists endeavour to show that the various agencies
engaged in degrading rocks and accumulating sediments
in former ages are not to be judged of by the agencies
we now see around us ; in former times they were more
active. At one period the elastic tides in the earth may
have been so great that they resulted in the fracturing
off from our planet its satellite the moon, and subsequently the moon, acting on the waters of the earth, may,
even as late as 150,000 years ago, have produced every
three hours tides 150 feet in height.

Whatever may be the value of the figures here quoted,
reasonings like these bring us to the conclusions that the
various agencies which we now know to be acting upon
our earth were once far more potent than they are at
present, and if the moon, as a producer of elastic tides,
has any influence upon the occurrence of earthquakes,

it must have had a much greater influence in bygone times.

We might speak similarly with regard to the internal heat of the earth.

From the present heat-gradient of our globe it is possible to calculate how much heat flows from the earth every year.

This is equivalent to a quantity which would raise a layer of water ·67 centimetres thick, covering the whole of our globe, from a temperature of 0° to 100° C.

Similarly, we might calculate the quantity of heat which would be lost when the average heat gradient, instead of being 1° F. for fifty feet of descent, was 1° F. for twenty-five feet of descent.

We might also calculate how many years ago it was since such a gradient existed.

The general result which we should arrive at would be that in past ages the loss of heat was more rapid than it is at present. Now the contraction of a body as it cools is for low temperatures proportional to its loss of heat, and this law is also probably true for contraction as it takes place from high temperatures.

Contraction being more rapid, it is probable that phenomena like elevations and depressions would be more rapid than they are at present, and generally all changes due to plutonic action, as has already been pointed out by Sir William Thomson, must have been more active.

We have, therefore, every reason to imagine that earthquakes which belong to the category of phenomena here referred to were also numerous and occurred on a grander scale during the earlier stages of the world's history than they do at present, and seismic and volcanic energy, when considered in reference to long periods of time, is probably a decreasing energy.

In making this statement we must not overlook the fact that in geological time, as testified by the records of our rocks, volcanic action, and with it probably seismic action, has been continually shifting, first appearing in one area and then in another, and even in the same area we have evidence to show that these have periods of activity and repose successively succeeding each other. Thus in Britain, during the Palæozoic times, we have many evidences of an intense volcanic activity. During the Mesozoic or Secondary period volcanic energy appears to have subsided, to wake up with renewed vigour in the Cainozoic or Tertiary period.

During this latter period it is not at all improbable that Scotland was in past times as remarkable for its earthquakes as Japan is at the present day.

Later on it will also be shown that earthquakes are concomitant phenomena, with those elevatory processes which we have reason to believe are slowly going on in certain portions of the earth's crust. If, therefore, we are able by the examination of the rocks which constitute the accessible portions of our globe to determine which periods were characterised by elevation, we may assume that such periods were also periods of seismic activity.

Amongst these periods we have those in which various mountain ranges appeared. Thus the Grampians, and the mountains of Scandinavia, were probably produced before the deposition of the Old Red sandstone. The Urals were upheaved prior to Permian times. The chief upheaval in the Alps took place after Eocene times. The Rigi and other sub-Alpine mountains were formed after the deposition of the Miocene beds. About this same time the Himalayas were upheaved.[1]

The earthquakes which from time to time shake those

[1] Ramsay, ' Geological History of Mountain Chains,' *Mining Journal.*

newer mountains apparently indicate that the process of mountain-making is hardly ended.

Seismic energy in relation to historical time.— The distribution of seismic energy with regard to historical time is a subject which has been very carefully examined by Mallet, who collected together a catalogue of between six and seven thousand earthquakes, embraced between the periods B.C. 1606 and A.D. 1850. The earthquake of B.C. 1606 was on the occasion of the delivery of the law at Mount Sinai. Between B.C. 1604 and B.C. 1586 an earthquake probably occurred in Arabia, when Korah, Dathan, and Abiram were swallowed up. Another biblical record is that of B.C. 1566, when the walls of Jericho were overthrown.

The earliest records from China is in B.C. 595; in Japan B.C. 285; in India A.D. 894.

By using the number of earthquakes which have been recorded in each century as ordinates, Mallet constructed a curve, which apparently shows a continual increase in seismic energy, especially during recent times. This, Mallet remarks, contradicts all the analogies of the physics of the globe, and points out that the rapid increase in the number of earthquakes in latter years is chiefly due to the greater number of records which have been made, and the increase of the area of observation. No doubt many of the records made by the ancients have been lost.

If we compare Mallet's records, as he invites us to do, with the great outlines of human progress, we see that the two increase simultaneously, and we come to the conclusion that, taken as a whole, during the historical period the seismic activity of the world has been tolerably constant.

These conclusions, based on the evidence at our

command, are not to be confuted.   If, however, instead
of considering the seismic energy of the whole world, we
consider the seismic energy of particular areas, it seems
reasonable to expect that in certain instances sometimes
a decrease and sometimes an increase in this energy
might be discovered, especially, perhaps, in areas which
are highly volcanic.

In France we know that volcanic activity ceased at a
period closely bordering on historical times, and it is not
unlikely that seismic activity may have ceased at a corresponding time.

In a country like Japan, it is possible that in one
district seismic energy may be on the increase, whilst in
another upon the decrease.

In a country like England, it is probable that the
seismic state is constant, and, whatever changes may be
now occurring, they are taking place at so slow a rate
that, even if our records of the historical period were
complete, we could hardly be expected to find these
changes sufficiently marked to be observable.

For purposes of reference, and also for examining the
present question, the table, page 240, has been compiled.
The earthquakes given are chiefly those which have been
recorded in histories as being more or less destructive.

In the second column of this table will be seen the
number of earthquakes which have occurred in Japan
during each century, the centuries being marked in the
first column.   In columns 3 to 18 inclusive are given the
number of earthquakes which have occurred during different centuries in the various countries and districts
mentioned at the head of each column.   These latter,
which are taken from the writings of Mallet, are given
for the sake of comparison with the Japanese earthquakes.
If we commence with the seventh century in the column

1	2	3	4	5	6	7	8	9	10	11	12	13	14	15	16	17	18	19
Centuries	Japan	Scandinavia and Iceland	British Isles and Northern Isles	Spanish Peninsula	France, Belgium, Holland	Rhine Basin	Switzerland and Rhine Basin	Danube Basin	Italy, Sicily, Sardinia, and Malta	Supplemental table for Italy, Sardinia, and Malta	Turco-Hellenic Territory, Syria, Ægean Isles, and Levant	United States and Canada	Mexico and Central America	Antilles	Cuba	Chili and La Plata Basin	Northern Zone of Asia	Approximate Intensity in the Kioto District of Japan
I	1																	
II																		
III	1																	
IV	1								6		23							
V	1								5		19							
VI	12				1				3		27							
VII	11				6				1		8							15
VIII	40								2	1	12							17
IX	17				21		19	19	6		7							60
X	20				2		2		3	3	5							24
XI	18		8	3	16		9		7	5	18							28
XII	16		11	4	12		8		18	22	23							20
XIII	19		15	3	9		3		15	26	13							16
XIV	36		4	8	21		18		20	51	8							25
XV	17		1	4	14		12		18	47	11							29
XVI	26		8	10	61	10	52	35	32	5	22		6	1	4	5		17
XVII	31	28	14	10	91	29	120	31	121	9	53	10	7	16	4	9		11
XVIII	31	111	63	93	237	71	141	88	438	20	124	88	24	85	2	10	32	8
XIX	27	113	110	85	211	81	173	145	390	88	194	51	30	145	50	170	57	8

for Japan, we see that a great increase in the number of earthquakes, as we come towards the present time, is not so observable as it is in the other columns.

The explanation for this probably lies in the fact that Japan has practised civilised arts for a longer period than many of the European and other countries mentioned in the table.

In Japan, no doubt, the records of later years have been more perfect than they were in early times, but this, although so potent an effacer of what was probably the true state of natural phenomena in the case of Europe, has not quite obliterated the truth in Japan; for instead of an apparent increase of seismic energy since early times it shows a slight decrease.

To draw up a table of earthquakes such as the one which has just been given, and then, after the inspection of it, draw conclusions as to whether there has been an increase or decrease in seismic energy, is, however, hardly a just method of reasoning. The earthquakes, taken as they are, for the whole of Japan, represent a collection of places some of which are 1,000 miles apart. When we consider that many earthquakes which occurred at one end of this line were never felt at the other end, in order to justly estimate the periodicity of seismic phenomena it would seem that we ought either to take some particular seismic area or else the whole world.

The particular area which has been taken is that of Kioto in Central Japan, and the earthquakes which have been felt there are enumerated in the table.

In order to show the variation in seismic activity of this district a curve has been plotted, fig. 35, with ordinates equal to the values given for the Kioto earthquakes during succeeding centuries. The upper points of these ascending and descending lines are

R

joined by a free curve. The lower points are similarly joined. The points of bisection of ordinates drawn between these two curves are taken as points in a curve to show the true secular change in seismic energy.

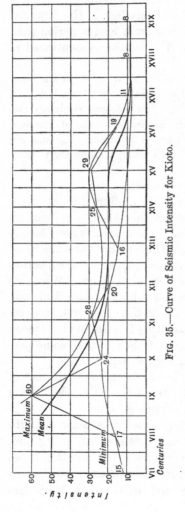

FIG. 35.—Curve of Seismic Intensity for Kioto.

By looking at this wavy line it will be seen that the intervals between maxima and minima are closer together in early times than they are later on.

Thus, between the eighth century and the ninth century, points of maximum and minimum seismic efforts occurred at times a century apart, whilst later on, from the eleventh to the fifteenth century, they were at intervals of 300 years apart.

By inspecting either the wavy line or the resultant curve, it will be seen that since the ninth century down to the present time there has been a decided decrease in seismic energy. From the ninth century down to the fifteenth century this decrease is represented by a regular curve. At this point, however, the decrease becomes slightly more rapid, and is represented by a second curve. If, instead of calculating ordinates for my curve, in which intensity has been considered, simply the number of earthquakes are counted, a similar result is obtained. From this it appears that the rate at which seismic energy decreased during the last 500 years was about the same as that at which it decreased during the 500 years previous to this period.

If the lists for the Italian and Turco-Hellenic districts could be similarly analysed, and the earthquakes of any particular district picked out from the others, it is very probable that a similar decrease or alteration in seismic energy might be observed.

Provided that we have at our disposal records of the various earthquakes which have occurred in any given district during a sufficiently long period of time, one conclusion that we may expect to arrive at is that we shall be able to trace some variation in the seismic activity of that district. For the Kioto area, it has been shown that there is a diminution in seismic activity. In other districts, however, there may possibly be an increase.[1]

Relative frequency of earthquakes.—A question which is of great interest to those who dwell in shaken districts is as to how often disturbances may be expected to occur.

From a general examination of this question, considering the earthquakes of the whole world, Mallet arrived at the following conclusions :—

[1] A notable example of a rapid diminution in the number of earthquakes felt at a place is that of Comrie in Scotland. In 1839–40, no less than sixty shocks were felt in eleven months. In 1842–43, about thirty shocks were felt, and in the following year thirty-seven. Since this time the number of shocks has decreased until they are almost of as rare occurrence at Comrie as in other portions of the British Isles.

1. While the smallest or minimum paroxysmal intervals may be a year or two, the average interval is from five to ten years of comparative repose.

2. The shorter intervals are in connection with periods of fewer earthquakes—not always with those of least intensity, but usually so.

3. The alternations of paroxysm and of repose appear to follow no absolute law deducible from these curves.

4. Two marked periods of extreme paroxysm are observable in each century, one greater than the other—that of greatest number and intensity occurring about the middle of each century, the other towards the end of each.

The form of the curves which Mallet has drawn seem to indicate that seismic energy came in sudden bursts, and then subsided, gradually gathering strength for another exhibition. This is continually seen in the shocks experienced in various seismic areas—a large shock, or the maximum of the activity dying out by repeated small shocks on succeeding days.

Mr. I. Hattori, writing on the large earthquakes of Japan, remarks that on the average there has been one large earthquake every ten years. They, however, occur in groups, as shown in the following table.

No. of shocks	Period	Interval
6	A.D. 827–836	10 years
6	„ 880–890	10 „
4	„ 1040–1043	4 „
5	„ 1493–1407	5 „
4	„ 1510–1513	4 „
5	„ 1645–1650	6 „
5	„ 1662–1664	3 „
4	„ 1853–1856	4 „

Dr. E. Naumann, who has also written on the earthquakes in Japan, remarks that if periods of seismic activity do not occur every 490 years, there is a repetition of the cycle after 980 years, but there is much variability. A period of 68 years is very marked. On the average, large earthquakes have occurred every 5·9 years. Fuchs gives some interesting examples of the repetition of earthquakes at definite intervals, of which the following are examples. Sometimes earthquakes appear to have repeated themselves after 100 years. One remarkable example of this is that of Lima, on June 17, 1578, which was repeated on the same day in the year 1678. In Copiapo it is believed that earthquakes occur every twenty-three years, and examples of such repetitions are found in the years 1773, 1796, and 1819. In Canada, near to Quebec, earthquakes lasting forty days are said to occur every twenty-five years. The plateau of Ardebil is said to be regularly shaken by earthquakes every two years.

A. Caldcleugh, writing on the earthquake of Chili, in 1835,[1] remarks that the Spaniards first had the idea that a great earthquake occurs every century. Afterwards they thought the period was every fifty years. As a matter of fact, however, there were large earthquakes in 1812, at Caraccas; in 1818, at Copiapo; in 1822, at Santiago; in 1827, at Bogota; in 1828, at Lima; in 1829, at Santiago; and in 1832, at Huasco.

The average period of seismic disturbances in any country probably depends upon the subterranean volcanic activity of that country. When the activity is great the large earthquakes may occur at short intervals; but when the activity is small, as in England, shocks of moderate intensity may not be felt more than once or twice per century. A general idea of the relative frequency of the

[1] *Phil. Trans.* vol. i. 1836.

large earthquakes in various parts of the world may be easily obtained by an inspection of the table on page 240.

Between the years 1850 and 1857 Kluge found that in the world there had been 4,620 earthquakes, which is, upon the average, nearly two per day. This estimate of the frequency of earthquakes of sufficient intensity to be recorded without the aid of instruments is, however, much below the truth. In Japan alone there probably occurs, as a daily average, a number at least equal to that which has been just given for the whole world. Boussingault considered that, in the Andes, earthquakes were occurring every instant of time.[1]

To state definitely how many earthquakes are felt in the world on the average every day is, from the data which we have at our command, an impossibility. Perhaps there may be ten, perhaps there may be 100. The question is one which remains to be decided by statistics which have yet to be compiled.

After a large earthquake, smaller shocks usually occur at short intervals. At first the succession of disturbances are separated from each other by perhaps only a few minutes or hours. Later on, the intensity of these shocks usually decreases, and the intervals between them become greater and greater, until, finally, after perhaps a few months, the seismic activity of the area assumes a quiescent state.

The great earthquake which overtook Concepcion on February 20, 1835, was followed by a succession of shocks like those just referred to, there being registered, between the date of the destructive shock and March 4, 300 smaller disturbances.

During the twenty-four hours succeeding the destruction of Lima (October 28, 1746), 200 shocks were

[1] *Am. Jour. of Sci.* vol. xxxvii. p. 1.

counted, and up to the 24th of February in the following
year 451 shocks were felt.

At St. Thomas, in 1868, 283 shocks were counted in
nine and a quarter hours.

Similar examples might be taken from the description
of almost all destructive earthquakes of which we have
records. For a large earthquake to occur, and not to be
accompanied by a train of succeeding earthquakes, is
exceptional. Sometimes we find that a large number of
small earthquakes have occurred without a large one
being felt. Seismic storms of this description have happened, even in England—for instance, in the year 1750,
which appears to have been a year of earthquakes for
many portions of the globe.

In this year, which is known as the 'earthquake year,'
shocks were felt in England as follows : On March 14, in
Surrey; March 18, in south-west of England ; April 2, at
Chester; June 7, at Norwich; August 23, in Lincoln-
shire ; September 30, Northamptonshire.

Synchronism of earthquakes.—One of the first writers
who drew attention to the fact that two shocks of earthquakes have been felt simultaneously at distant places
was David Milne, who published a list of these occurrences.[1]

In two instances, February and March 1750, shocks
were simultaneously felt in England and Italy. In
September 1833 shocks appear to have been simultaneously felt in England and Peru. These and many
other similar examples are discussed by Mallet, who
thinks with Milne that these coincidences are in every
probability matters of accident. According to Fuchs,
Calabria and Sicily appear often to have had earthquakes
at the same time, as for instance in 1169, 1535, 1638,

---

[1] Milne, ' British Earthquakes,' *Edin. Phil. Journ.* vol. xxxi.

when the town Euphemia sank, and in the years 1770, 1776, 1780, and 1783.

A remarkable example of coincidence occurred on November 16, 1827, when a terrible earthquake was felt in Columbia, and at the same time a shock occurred on the Ochotsk plains, nearly antipodal to each other.

Kluge also gives a large number of instances of simultaneous earthquakes; thus, on January 23, 1855, on the same day that Wellington, New Zealand, so severely suffered, there was a heavy earthquake in the Sieben-geberge, and also in North America. To this might be added the fact that the last destructive earthquake in Japan occurred within a few days of this time.

Sometimes neighbouring countries where earthquakes are common are equally remarkable by their utter want of synchronism. For example, Southern Italy and Syria are said never to be shaken simultaneously.

Secondary earthquakes.—Although it is possible that the simultaneous occurrence of earthquakes in distant regions may sometimes be a matter of chance, it must also be remarked that the shaking produced by one earth-quake may be sufficient to cause ground which is in a critical state to give way, and thus the first disturbance becomes the originator of a second earthquake. Admit-ting that an earthquake, as it radiates from its centre, may act in such a manner, we see that a feeble disturb-ance might be the ultimate cause in the production of a destructive earthquake, just as the disturbance of a stone upon the face of a scarp might, by its impact upon other stones, cause many tons of material to be dislodged.

It is also easy to conceive how the seismic activity of two districts may be dependent upon each other. Inas-much as these secondary shocks are direct effects of primary disturbances, they might have been treated in a previous chapter.

As examples of consequent or secondary earthquakes Fuchs tells us that when small earthquakes take place in Constantinople and Asia Minor, earthquakes are felt in Bukharest, Galazy, and Kronstadt.

The great Lisbon earthquake also appears to have given rise to several consequent disturbances. One was in Derbyshire, occurring at 11 a.m. It was sufficiently violent to cause plaster to fall from the sides of a room and a chasm to open on the surface of the ground. Some miners working underground were so alarmed that they endeavoured to escape to the surface. During twenty minutes there were three distinct disturbances.

Another shock was felt at Cork.[1]

Although these disturbances own a consequence of the Lisbon earthquake they might properly perhaps be attributed to the pulsations produced by the shock at Lisbon, which spread through England and other countries without being felt.

The shocks which men felt in New Zealand and New South Wales in 1868 were probably secondary shocks, due to the disturbance at Arequipa and other places on the South American coast.

These so-called secondary earthquakes, although in many instances they may be due to earth pulsations produced by earthquake, or to the immediate sensible shaking of a large earthquake, may perhaps, in certain instances, be attributed to some widespread disturbance beneath the crust of the earth. The occurrence of periods where all earthquake countries suffer, unusual disturbances indicate the probability of such underground phenomena.

[1] *Phil. Trans.* vol. xlix. pt. i.

CHAPTER XIV.

DISTRIBUTION OF EARTHQUAKES IN TIME (*continued*).

The occurrence of earthquakes in relation to the position of the heavenly
bodies—Earthquakes and the moon—Earthquakes and the sun; and
the seasons; the months—Planets and meteors—Hours at which
earthquakes are frequent—Earthquakes and sun spots—Earthquakes
and the aurora.

*The position of the heavenly bodies and the oc-
currence of earthquakes.*—Since the earliest times, in
searching for the cause of various natural phenomena,
man has turned his energies towards the heavens.   One
of the earliest observations was the connection that exists
between the season of the year and the motions of the
heavenly bodies.   Tides were seen to be influenced by the
moon.   In later times it has been discovered that periods
of maximum magnetic disturbances occur every ten or
eleven years with the sun spots, and Herr Kreil, of Vienna,
tells us our satellite, the moon, has also an influence
upon the magnet.

From day to day we see the bond connecting our
planet with the sun, the moon, and other heavenly bodies
which are outside us gradually becoming closer.

Inasmuch as many phenomena, like the motion of the
tides, the rise and fall of the barometer, fluctuations in
temperature, are all more or less directly connected with
the relative position of our planet with regard to the sun

and moon, any coincidence between the phases of these bodies and the occurrence of earthquakes more or less involves a time relationship with the other phenomena resultant on lunar and solar influences.

Earthquakes and the position of the moon.—Many earthquake investigators have attempted to show the connection between earthquakes and the phases of the moon.

The first and most successful worker in this branch of seismology was Professor Alexis Perrey, of Dijon, who, after many years of arduous labour in tabulating and examining catalogues of earthquakes, showed that earthquakes were more likely to occur at the following periods than at others.

1. They are more frequent at new or full moon (syzygies) than at half moon (quadratures).

2. They are more frequent when the moon is nearest the earth (perigee) than when she is farthest off (apogee).

3. They are more frequent when the moon is on the meridian than when she is on the horizon.

These results were obtained by Perrey after analysing his catalogues by three different and independent methods, and they were confirmed by the report of a committee appointed by the Academy of Sciences. It must, however, be remarked that in several instances anomalies occur, and also that the difference between the number of earthquakes at any two periods is not a very large one. Thus, for instance, the annual catalogues compiled by Perrey from 1844 to 1847, the earthquakes in perigee are to those in apogee as 47 : 39. Between the years 1843 and 1872 Perrey finds that 3,290 shocks occurred at the moon's perigee, and 3,015 at the apogee.[1]

[1] *Compte Rendus*, 1875, p. 690.

Between 1761 and 1800 earthquakes occurred as
follows :—

In Perigee 526
Apogee 465

The following table shows the results which enabled
Perrey to deduce his first law.

Dividing the period of lunation into quarters, with
the time of syzygies and quadratures as the centres of these
quarters, he found that the earthquakes were distributed
as follows.

	Totals	Syzygies	Quadratures	Difference in favour of the Syzygies
1843–1847	1,604	850·48	753·52	69·96
1848–1852	2,049	1,053·53	995·47	58·06
1853–1857	3,018	1,534·13	1,483·87	50·26
1858–1862	3,140·	1,602·99	1,537·41	65·98
1863–1867	2,845	1,463·42	1,381·58	81·84
1868–1872	4,593	2,333·48	2,259·52	73·96
1843–1872	17,249	8,838·03	8,410·97	427·06

The reported earthquakes between 1751 and 1843 are
shown to conform with the same rule.[1] Julius Schmidt,
astronomer at Athens, found for the earthquakes of
Eastern Europe and adjacent countries for the years 1776
to 1873 that there were more earthquakes when the moon
was in perigee. Other maxima were at new moon, and
two days after the first quarter. There was a diminution
at full moon, and a minimum on the day of the last
quarter. As one example of results which are antagonistic
to the general results obtained by Perrey may be quoted
the results of an examination by Professor W. S. Chaplin
of the earthquake recorded at the meteorological observatory in Tokio. The list of earthquakes, 143 in number,
extending over a period of three years, was recorded by one
of Palmieri's instruments. The results were as follows :—

[1] *Am. Jour. Sci.* vol. xi. p. 233.

1. There have been maxima of earthquakes when the moon was two and nine hours east and seven hours west. At the upper transit there is a minimum.

2. Considering the moon's position with regard to the sun, at conjunction there were 32, at opposition 37, and at quadrature 74. East of the meridian the maximum was at least four hours.

3. When the moon was north of the equator these were 68, when south 82.

4. A maximum of earthquakes seven and eleven days after the moon's perigee. The fact that these results were obtained for the earthquakes of a special small seismic area renders them more interesting.[1]

Frequency of earthquakes in relation to the position of the sun.—The question as to whether there is a connection between the frequency of earthquakes and the relative position of the sun is to a great extent identical with the question as to the relative frequency of earthquakes in the various seasons. It is a subject which we find referred to by writers in the earliest ages. Pliny and Aristotle thought that earthquakes occurred chiefly in spring and autumn. In later times it has been a subject which has been most carefully considered by Merian, von Hoff, Perrey, Mallet, Volger, Kluge, and others who have devoted attention to seismology. In a résumé of the earthquakes of Europe, and of the adjacent parts of Asia and Africa, from A.D. 306–1843, Mallet gives the following results :—

	For Nineteenth Century		For the whole period	
Winter Solstice .	177	Solstices	253	Solstices
Spring Equinox .	151	306	170	403
Summer Solstice .	129	Equinoxes	150	Equinoxes
Autumnal Equinox	164	315	159	329

[1] *Transactions of the Asiatic Society of Japan*, vol. vi. pt. i. p. 353.

The above periods were called by Perrey *critical epochs*, because as a general result of his researches be found that at such periods there was a greater frequency of earthquakes. Fuchs, quoting from Kluge's tables, extending from 1850–1857, tells us that the recorded earthquakes occurred as follows:—

In the Northern Hemisphere—

Equinoxes	**1324**
Solstices	1202

In the Southern Hemisphere—

Equinoxes	**301**
Solstices	261

Earthquakes are, therefore, more frequent at the equinoxes, and this especially at the autumnal equinox. In the northern hemisphere, at the solstices, the greater number of shocks occur about the winter solstices, whilst in the southern hemisphere, about the summer solstices.

ᵣExceptions, however, are found in Central America and the West Indies, in the Caucasus, and the Ægean Sea.

The idea that earthquakes had a periodicity dependent upon the position of the heavenly bodies is by no means confined to Europe. In a Japanese work called ' Jishin Setsu' (an opinion about earthquakes) by a priest called Tensho, it is stated that the relative positions and movements of the twenty-eight constellations with respect to the moon cause earthquakes. This Tensho asserts after careful calculation, and Falb tells us that all future earthquakes can be predicted.

In the Kuriles and Kamschatka, Sicily, and in parts of South America, it is said that the equinoxes are regarded as dangerous seasons.

Frequency of earthquakes in relation to the seasons and months.—What is here said respecting the relative frequency of earthquakes at the different seasons and months is little more than an extension and critical examination of the results which have been given respecting the frequency of earthquakes in regard to the position of the sun.

That there is a difference between the number of earthquakes which are felt at one season of the year as compared with those felt at another is a fact which, as seismoscopic observations are extended, is becoming more and more recognised.

Some of the more important results which were arrived at by Mallet from 5,879 observations made in the northern hemisphere, and 223 in the southern hemisphere, may be expressed as follows :—

	Maxima	Minima
Northern Hemisphere .	January, also a slight rise in August and October	May, June, and July
Southern Hemisphere .	November, also May and June	March, extending over one month, also August . .

Julius Schmidt, of Athens, who so carefully examined the earthquakes of eastern Europe, came to the following conclusions :—

For the earthquakes between 1200 and 1873, a maximum on September 26 and January 17 ; a minimum on December 3 and June 13.

For the earthquakes between 1873 and 1874, a maximum on March 1 and October 1 ; a minimum on July 7 and December 15.

For all the earthquakes of eastern Europe, a maximum on January 3 ; a minimum on July 8, or there was a maximum at perihelion and aphelion.

When the months are grouped together according to the seasons, spring, summer, autumn, and winter, we find that in the northern hemisphere the minimum is in summer and the maximum in winter, whilst in the southern hemisphere (giving the proper months corresponding to its seasons) we find two maxima, one at the commencement of winter, and the other at midsummer, whilst the minima are in spring and autumn.

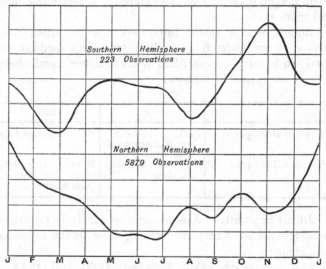

FIG. 36.—Curves of Monthly Seismic Intensity (Mallet).

In the following table the difference in the number of earthquakes felt at different seasons is given more in detail.

In examining this table, we must remember that for countries like Peru, Chili, and New Zealand, lying in the southern hemisphere, the records given for the months April to September correspond to the winter months of those countries. The Roman numerals indicate the centuries between which the records date.

		October to March	April to September
Northern Regions	1. Scandinavia and Iceland, xii–xix .	129	91
	2. British and Northern Isles, xi–xix .	123	94
	3. Belgium, France, and Holland, iv–xix	395	272
	4. Rhone Basin, xvi–xix . . .	115	69
	5. Switzerland and Rhine Basin, ix–xix	327	205
	6. Danube Basin, v–xix . . .	147	128
	7. Spanish Peninsula, xi–xiv . .	114	87
	8. Italy, Sicily, Sardinia, and Malta, iv–xix	650	581
	9. Turco-Hellenic Territory, Syria, Ægean Isles, and Levant, iv-xix .	214	222
	10. Northern Zone of Asia, xviii–xix .	46	36
	11. Japan (Tokio area), 1872–1880 (small earthquakes) . .	213	157
	12. Japan B.C. 295–A.D. 1872 (large earthquakes)	165	188
	13. Algeria and Northern Africa . .	26	20
	14. United States and Canada, xvii–xix	86	48
Central Regions	15. Java, Sumatra, and neighbouring Islands, 1873-4-7-8 . . .	194	182
	16. Mexico and Central America, xvi–xix	26	26
	17. West Indies (Mallet), xvi–xix . .	108	114
	18. West Indies, xvi–xix . . .	296	343
	19. Cuba, xvi–xix	28	23
Southern Regions	20. Chili, and La Plata Basin, xvi–xix .	89	89
	21. Peru, Columbia, Basin of Amazons, xvi–xix	506	541
	22. New Zealand, 1869–1879 . . .	166	176

Neglecting those records which show as many earthquakes for the winter months as for the summer months, we see at a glance that generally the greater number of shocks have happened during the colder seasons. In the southern hemisphere, so far as the records go, this is not true. In the northern regions, out of fourteen examples there are two exceptions. In the central regions there are two cases where the greatest number of earthquakes have been recorded in the winter months, and two cases where the greatest number have been recorded for the summer.

Altogether, out of twenty-two examples, there are only six exceptions to the rule. These exceptions altogether occur among records many of which are ancient, and are, therefore, more open to error than lists which have been compiled in modern times.

Because small earthquakes are seldom noticed by persons out in the open air, it might be expected that the number of earthquakes observed in warm countries at one portion of the year would be equal to those observed in any other season. Such an argument, however, would hardly apply to most of the records which are quoted, as they refer to destructive disturbances.

If, however, we take the records made in tropical countries from the table just given, we see that in such countries there have been almost as many observations of earthquakes at one season as at any other.

Another fact which might be adduced against the rule that the greater number of earthquakes occur during the winter months would be the comparison of a table of earthquakes recorded previous to the nineteenth century. By doing this we see that for certain countries the winter rule is inverted, and that the greater number of shocks are felt during the summer.

Notwithstanding these objections to Perrey's conclusions, the balance of evidence is in favour of his general result, and we may conclude that during the colder portions of the year we may expect more shakings than during the warmer portions. Comparing the number of earthquakes of winter and autumn to those of summer and spring, they are to each other in the proportion of 4 : 3.

A fairer way to examine this question, and to determine what is probably the present state of seismic activity in our globe, would be only to consider the earthquakes which have taken place in comparatively recent times, laying especial stress upon those observations which have been made with the assistance of automatic instruments, or those which have been collected by persons interested in these investigations.

For this purpose the following table, showing the distribution of earthquakes in different countries during the nineteenth century, has been compiled.

The arrangement is mensual. Where the number of earthquakes in any month is above the average, the number is printed in large type; where below the average, in small type.

EARTHQUAKES OF THE NINETEENTH CENTURY, CHIEFLY FROM PERREY.

	January	February	March	April	May	June	July	August	September	October	November	December	Average per month
Scandinavia and Iceland .	17	11	11	7	7	6	8	8	10	10	11	6	9·3
British Isles and Northern Isles	9	9	10	7	8	6	5	11	12	8	11	12	9
France, Belgium, Holland .	27	17	21	13	13	8	15	17	15	17	21	25	17
Basin of the Rhone . .	12	12	8	3	3	2	2	4	6	6	8	14	6·6
Basin of the Rhine and Switzerland . . .	15	17	13	12	11	6	12	11	10	17	24	25	14
Basin of the Danube . .	14	15	9	8	12	8	16	11	11	16	10	12	11·8
Spanish Peninsula . .	10	5	6	7	4	6	10	5	9	11	7	5	7
Italian Peninsula, Sicily, Sardinia, and Malta . .	44	44	48	43	40	34	41	46	27	45	26	39	39
Turco-Hellenic Territory, Syria, Ægean Islands, and Levant	22	20	10	10	16	15	14	22	14	17	12	14	16
Northern Zone of Asia .	4	6	6	4	4	3	5	7	6	3	4	5	4·7
1876-1881, Japan (Tokio area)	39	41	41	30	33	30	27	21	10	28	34	43	31·4
Japan (large earthquakes)	—	5	3	3	1	—	5	4	—	—	1	5	2
Algeria and North'rn Africa	5	2	6	7	3	2	2	5	1	4	8	1	3·8
United States and Canada .	4	4	3	3	3	—	4	6	3	2	7	5	3·8
Java, Sumatra, &c., 1873-4-5-7 and 9	35	30	38	33	22	36	27	40	24	35	30	26	31
Mexico and Central America	3	2	2	2	6	2	2	1	1	.3	2	3	2·5
Antilles	9	8	19	12	12	10	9	16	12	10	13	12	11·8
Cuba . . . . :	4	3	2	3	3	4	5	2	6	5	6	4	4
Chili and La Plata . . :	14	10	14	8	19	11	16	16	9	27	8		13·9
Peru, Columbia, Basins of Amazons, xvi–xix . .	92	83	92	27	106	79	94	93	97	77	72	90	87
New Zealand, 1869–79 .	31	27	37	23	22	31	27	36	37	21	27	23	28·5
Jan. 1850, Dec. 1857 Northern Hemisphere .	153	162	143	161	126	124	141	156	154	171	151	168	15·0
Southern Hemisphere .	75	43	61	66	46	42	53	39	54	55	57	46	53
1821–1830 Northern Hemisphere .	31	36	31	29	33	33	20	31	24	41	26	34	30
Southern Hemisphere .	2	—	1	1	3	1	3	2	3	2	1	1	1·6

A glance at this table shows that for most countries in the northern hemisphere the rule that there are generally more earthquakes during the winter months—that is, from October to March—holds good. For countries which lie comparatively near to the Equator, and also for those countries in the southern hemisphere, the rule is not so clear. When examining this table it must be remembered that it does not enable us to judge of the relative frequency of earthquakes in different countries, inasmuch as the periods over which the records were taken are different in different cases.

To the above table might be added the records of P. Merian, who examined the earthquakes felt in Basle up to 1831. As a result he found that during the winter months eighty shocks had been felt, whilst during the summer only forty. Taking the records for the two hemispheres from 1850–1857, compiled by Kluge,[1] in the northern hemisphere we have in the months between October and March 948 shocks against 862 in the remainder of the year. In the same months in the southern hemisphere we have for the corresponding periods the numbers 337 and 300, and thus both hemispheres would appear to follow the same rule. If, however, we examine the table we see that the two seasons are not so pronounced for the southern hemisphere as they are for the northern, and that there may be two or three periods of maximum disturbance as has been previously indicated.

Earthquakes and the planets and meteors.—Just as the moon and the sun may exert an attractive influence upon the earth and cause earthquakes to predominate at certain seasons rather than at others, several investigators of seismic phenomena have thought that the planets might act in a similar manner.

[1] Kluge, *Ueber die Ursachen*, &c., p. 74.

M. J. Delauney, from a study of Perrey's tables of earthquakes from 1750–1842, found two groups of maxima each with a period of about twelve years, one commencing in 1759 and the other in 1756. Two other groups with twenty-eight year periods respectively commence in 1756 and 1773. These groups coincide with the times when Jupiter and Saturn reach the mean longitudes of 265° and 135°. From this Delauney concludes that earthquakes have a maximum when the planets are in· the mean longitudes just mentioned.

The increased number of earthquakes, especially in November, are attributed to the passage of the earth through swarms of meteors, and in like manner supposes the influence of Jupiter and Saturn to be due to their passing through meteor streams situated in mean longitudes 135° and 265°.

As a consequence of this he predicts an increase of earthquakes in the years 1886, 1891, 1898, 1900, &c.[1]

Dr. E. Naumann, who critically examined the large earthquakes of Japan, showed that there was an approximate coincidence between many of the disturbances and the thirty-three year period of meteoric showers.[2]

Humboldt states that a great shower of meteors was seen at Quito before the great earthquake of Riobamba (Feb. 4, 1797). The earthquakes of 1766 and 1799 at Cumana are also said to have been accompanied with meteoric showers. Mallet gives a list of large earthquakes which occurred at the times when meteors were observed.[3]

The hours at which earthquakes are most frequent.— From the examination of a catalogue of over 2,000 earth-

[1] *Am. Jour. Sci.* vol. xix. p. 162.
[2] *Mitth. d. Deutsch. Ges.*, Aug. 1878.
[3] *Report to British Association*, 1850, p. 74.

quakes which occurred in various parts of the world
between the years 1850 and 1857, made by Kluge, it is
found that both for the northern and southern hemispheres the observations which were made during the night
generally exceed those which were made during the day.

	Number of Earthquakes	
	Day	Night
In the Northern Hemisphere . . .	938	1592
In the Southern Hemisphere . . .	292	357

In the northern hemisphere the greatest number were
observed between 10 P.M. and 12 P.M. (360 shocks), and
the fewest between 12 and 2 P.M. (139 shocks). In the
southern hemisphere, the greatest number were observed
at night between 12 and 1, and the smallest number
between 1 and 2 and 4 and 5 in the afternoon.[1] These
distinctions, however, are less distinctly marked as we
approach the Equator. Schmidt found for the earthquakes of the Orient between 1774 and 1873, that shocks
had been most frequent about half-past two A.M., and less
frequent about 1 P.M. With regard to these conclusions,
which have been reached with much labour, we might be
inclined to think that they are partially to be explained
on the supposition that more observations are made during
the night than during the day—the personal experience
of residents in an earthquake country being, that many
earthquakes which occur during the day are passed by
unnoticed, whilst those which occur during the night are
recorded by thousands of observers. Such a view is certainly confirmed by the instrumental records obtained in
Japan. From 1872 to 1880 inclusive there were 261
shocks recorded, 132 of which occurred between the hours
of 6 P.M. and 6 A.M.

[1] Fuchs, *Die Vulkanischen Erscheinungen der Erde*, p. 424.

Earthquakes and sun spots.—Of late years considerable attention has been drawn to a coincidence between the occurrence of sun spots, magnetic disturbances, rainfall, and other natural phenomena.

These periods of sun spots occur about every eleven years, and appear to be coincident with the periodical return of the planet Jupiter. In Japan, Dr. E. Naumann sought for a coincidence between these periods of sun spots and earthquakes, but without any marked results.

Schmidt, who carefully compared his lists of earthquakes with the appearance of sun spots, came to the conclusion that there was no marked coincidence. The occurrence of earthquakes had sometimes synchronised with sun spots, whilst at other times there had been a maximum of sun spots and no earthquakes.

M. R. Wolf[1] apparently considers that earthquakes, like volcanic eruptions and the appearance of the aurora, are coincident with sun spots.

Kluge, however, came to the conclusion that when there are few sun spots, earthquakes, like volcanic eruptions and magnetic disturbances, have been at a maximum.

M. A. Poey, who examined a catalogue of the earthquakes of Mexico and the Antilles, extending from 1634 to 1870, shows by a table that earthquakes have come in groups, first at the maxima and then at the minima period of sun spots. Out of thirty-eight groups, seventeen being at the maximum and seventeen at the minimum, the remaining four are exceptions to the rule, being between the maximum and minimum. Phenomena which are dependent upon heat occur with the minima of sun spots, and those dependent upon cold with the maxima.[2]

---

[1] *Bern. Naturf. Gesellschaft*, 1852.
[2] *Comptes Rendus*, 1874, Jan. to June, p. 51.

Earthquakes and the aurora.—The possible connection between earthquakes and the aurora is a subject which has attracted some attention. Boué has especially made a careful examination of this subject.[1]

He comes to the conclusion that if we compare the monthly periods of earthquake frequency and the aurora there is an agreement between the two. Comparing Perrey's tables of earthquakes from the fourth to the nineteenth century, with tables of the aurora, one-third of both phenomena have occurred, not only in the same day, but often at the same hour. Between 1834 and 1847, 457 earthquakes are given and 351 notices of the aurora.

Out of these :—

48 occur on the same day,
5 occur in the same hour,
30 approximate to the same time.

The nearer together that these phenomena have occurred the stronger have they been.

Professor M. S. di Rossi brings forward many examples where there has been a coincidence between the appearance of the aurora and earthquakes. On 139 nights out of 211 days the aurora was seen in some parts of Italy, and ninety-three times earthquakes were felt. On forty-six occasions earthquakes and aurora took place together.[2] In considering the probability of a connection existing between these two phenomena, we must bear in mind that the aurora is at no great height above the surface of our earth, and, further, that it can be partially imitated. The fact that in earthquake countries, like Japan, the aurora is practically never seen, would indicate that

---

[1] Boué, *Parallele der Erdbeben, Nordlichter und Erdmagnetismus, in Sitz. der K. A. d. Wissensch.* 1856, vol. iv. p. 395.

[2] *Meteorologia Endogena*, vol. i. p. 107, &c.

we can neither regard this imperfectly understood phenomenon either as an effect or cause of earthquakes. That earthquakes and the appearance of the aurora in certain countries should not sometimes coincide is an impossibility.

Dr. Stukeley, who, it must be remembered, attempted to correlate the phenomena of earthquakes and electricity, when writing of the disturbances which shook England in 1849 and 1850, says that the weather had been unusually warm, the aurora borealis frequent and of unusually bright colours, whilst the whole year was remarkable for its fire-balls, lightnings, and corruscations.[1]

The aurora was observed before the commencement of the Maestricht earthquakes in 1751 [2]; whilst at the time of the shock flashes of light like lightning were observed in the sky.

Glimmering lights were seen in the sky before the New England earthquakes (Nov. 18, 1755), and again, before the disturbances which occurred in the same region in 1727, peculiar flashes of light were seen.

Preceding the Sicilian earthquake of 1692 strange lights were seen in the sky. Ignis fatui have also been observed with earthquakes. At the time of auroral displays Bertelli has observed microseismical disturbances, and M. S. di Rossi, who has made similar observations, thinks that there is an intimate connection between the aurora and earthquakes; the aurora either occurring in a period of earthquakes, or else taking the place of earthquakes.

[1] *Phil. Trans.* vol. lxviii. p. 221.   [2] *Gent. Mag.* vol. xxvii. p. 508.

CHAPTER XV.

BAROMETRICAL FLUCTUATIONS AND EARTHQUAKES—FLUCTUA-
TIONS IN TEMPERATURE AND EARTHQUAKES.

Changes in the barometer and earthquakes.—Mallet,
who collected together a number of examples of earthquakes which have occurred with a fall of the barometer,
and a number which have happened with a rise, concludes
that there are as many instances of the one as of the
other. At the great earthquake of Calabria, in 1783, the
barometer was very low. The earthquake of the Rhine
(February 23, 1828) was preceded by a gradual fall of the
barometer, which reached its lowest point upon that day.
After the earthquake the barometer again rose. The earthquake of February 22, 1880, in Japan, was accompanied
by exactly similar phenomena. Caldcleugh, who observed
the heavy shocks in Chili (February 20, 1835), noticed that
on February 17 and 18 the barometer fell 5/10 inches.
Similar phenomena were observed before the succeeding
smaller shocks. After the shocks the barometer again
rose. Principal Dawson, speaking of the earthquakes of
Canada, observes that some of the shocks have been
accompanied with a low barometer.

P. Merian, who examined the connection between the
Swiss earthquakes and atmospheric pressure, found that
out of twenty-two earthquakes observed in Basle between

1755 and 1836, thirteen of these were local shocks, of which eight were accompanied with sudden changes of pressure. Of the remaining nine, which were only felt slightly in Basle, no change in atmospheric pressure was observed. Of thirty-six earthquakes which, between 1826 and 1836, were felt in Switzerland, thirty were chiefly confined to Switzerland, and ten of these occurred with a low or falling barometer.

Humboldt is of opinion that earthquakes only occur with changes in barometric pressure in those countries where earthquakes are few; and he gives examples where the regular variations of the barometer have gone on without interruption at the time of earthquakes.

Frederick Hoffmann, who examined fifty-seven earthquakes which occurred at Palermo between 1788 and 1838, came to the following result :—

The barometer was sinking	.	.	.	in 20 cases
„	„	rising .	. . .	in 16 „
„	„	at a minimum .	. .	in 7 „
„	„	„ maximum .	. .	in 3 „
„	„	undetermined .	. .	in 11 „ [1]

The observations of M. S. di Rossi apparently show that the earthquakes in Italy chiefly occur with a barometrical depression and with sudden jumps in atmospheric pressure.

Schmidt, who examined the earthquakes of the Orient, which occurred between 1858 and 1873, says that they were rare with a high barometer, but numerous when the barometer was low.

From an examination of a table of 396 earthquakes (May 8, 1875—Dec. 1881) felt in Tokio, furnished to me by Mr. Arai Ikunosuke, the director of the meteorological department, I obtained the following results :—

[1] *Die Vulkanischen Erscheinungen der Erde*, p. 419

The barometer was rising   .   .   .   .   in 169 cases
        „   „   falling   .   .   .   .   in 154   „
        „   „   steady   .   .   .   .   in 73   „
        „   „   below the monthly mean .   in 189   „
        „   „   above   „   .   in 192   „

From this it would appear that in Japan at least the
movements of the barometer do not show any marked
connection with the occurrence of earthquakes.

When considering this question we must remember
the marked effects which a lowering of the barometer
produces upon certain volcanoes and solfataras.   The
volumes of steam emitted from Stromboli and from some
of the solfataras in Tuscany hold a marked connection
with atmospheric pressure as the quantity of fire damp
given off from coal seams—these being greatest when the
barometer is low.   At certain changes of the weather it is
said that the volcano of Vulture, near Melfi, emits noises.
These phenomena at once place volcanic phenomena and
barometrical pressure in direct relationship.

Changes in temperature.—If, with an earthquake, it
should happen that there is a change in the height of
the barometer, we should naturally expect that this might
be accompanied with the changes in the temperature,
in the wind, and in other atmospheric phenomena which
are more or less connected with the height of the
barometer.

Many times it has been observed that after an earthquake there has been a sudden fall in the temperature.
Such was the case with the Yokohama earthquakes of
1880.

Cotte endeavours to show that the earthquakes of
Lisbon produced a change upon the temperature of all
Europe.   In the year which followed this earthquake
storms were more common than usual.

Kluge has collected together a large number of

examples when there has been a fall of temperature at the time of an earthquake.[1]

At Kiachta, in Siberia, at the time of the earthquake of December 27, 1856, the thermometer fell from 12° to 25° R.   We must, however, remember that there are many cases known where the thermometer rose.

M. S. di Rossi remarks that we have the highest records of temperature in the years richest in earthquakes. Thus, in 1873, at the time of the earthquakes in Central and Northern Italy, an abnormal high temperature was remarked.   Japanese writers have remarked upon the unusual heat which has shaken their countries.   The temperature of subterranean waters have been known to increase before earthquakes.

[1] Petermann's *Geogr. Mitth.* 1858, sec. 246.

CHAPTER XVI.

RELATION OF SEISMIC TO VOLCANIC PHENOMENA.

Want of synchronism between earthquakes and volcanic eruptions—
Synchronism between earthquakes and volcanic eruptions—Conclusion.

*Connection between earthquakes and volcanic erup-
tions.*—Insomuch as it is a recognised fact that regions
which are characterised by their seismic activity are
chiefly those which are also characterised by the number
of their volcanoes, it is generally assumed that these two
phenomena have an intimate relation. The residents in
a volcanic country, when seeking for the origin of an
earthquake, invariably turn towards the volcanoes which
surround them. If a neighbouring volcano is in a state
of activity, it is often regarded as a safeguard against
seismic convulsions, in other cases it is looked upon as
being the cause of such disturbances. In certain instances both of these views have apparently been corro-
borated. When we consider that an earthquake and a
volcanic eruption may both be the result of some great
internal convulsion, and that first one and then the other
may take place in the same neighbourhood, it is natural
to expect that when these internal forces have expended
themselves in the production of one of these phenomena,
it is not so likely that they should exhibit themselves in
the other. The inhabitants of Sicily and Naples, we are

told, regard eruptions of Etna and Vesuvius as safeguards against earthquakes. A similar belief is to be found in portions of South America with regard to the volcanoes for which that country is so celebrated.

From an examination of the records of the large earthquakes and the volcanic eruptions which have taken place in Japan during the last 2,000 years, Dr. Naumann found that there was often an approximate coincidence between the times of the occurrence of these phenomena, suggesting the idea that the efforts which had been sufficient to establish the volcano had at the same time been sufficient to shake the ground.

Of destructive earthquakes which have occurred at the time of volcanic eruptions, and of examples when these phenomena have occurred at widely separated intervals, the records are extremely numerous.

Want of synchronism between earthquakes and volcanic eruptions.—Many of the great earthquakes of South America do not appear to have been connected with volcanic eruptions.

The great earthquakes of the world, like those of Calabria and Lisbon, which took place in regions which are not volcanic, have not, Fuchs tells us, taken place in conjunction with volcanic outbursts.

In Japan, as in the Sandwich Islands and in many other parts of the globe, the small earthquakes which occur almost daily do not appear to show any marked connection with volcanic disturbances.

In 1881, during the eruption of Natustake, a volcano lying about a hundred miles north of Tokio, there was neither an increase nor a decrease in the earthquakes which were felt in Tokio. Similar remarks apply to the state of seismic activity of 1876–77, when Oshima, a volcanic island about seventy miles to the south of Tokio, was in eruption.

In the Sandwich Islands Mauna Loa seems to have its
eruptions independently of the disturbances which shake
these islands.[1]

*Synchronism of earthquakes and volcanic erup-
tions.*—Although many examples like the above may be
quoted, which apparently show an utter want of connection
between earthquakes and volcanoes, we must not overlook that class of earthquakes which almost invariably
accompany all great volcanic disturbances. In fact the
sudden explosions which take place at volcanic foci, as, for
instance, at the commencement of an eruption, are enumerated as one of the causes which produce earthquakes.
Earthquakes like these usually continue until the pressure of
the steam and lava have found for themselves an opening.
As compared with the total number of earthquakes which
are recorded, they form but an insignificant portion.

The direct connection which exists between these
phenomena has, no doubt, done very much to spread
the popular belief that all earthquakes may be connected
with volcanic eruptions.   As examples where this connection has existed we might quote from almost all the
volcanic countries in the world.

Thus, Fuchs tells us that on October 6, 1737, almost
the whole of Kamschatka and the Kurile Islands were
disturbed by movements which were simultaneous with
the outbreak of the great volcano Klutschenskja of North
Kamschatka.

One of the earliest records of a severe earthquake and
a volcanic eruption occurring simultaneously is found
in the accounts of the destruction of Herculaneum and
Pompeii.   The throwing up of Monte Nuovo in the neighbourhood of Pozzuoli was accompanied with a dreadful
earthquake.[2]

---

[1] *Notes on volcanoes of the Hawaiian Islands*, W. T. Brigham, Mem.
Boston Soc. of Nat. Hist., 1868.   [2] *Gent. Mag.* vol. xxiii., 1753.

In 1868 the earthquake of Arequipa was accompanied by the opening of the volcano Misti, on its north side. The distance to the volcano is about fourteen miles.

At the time of the eruptions of Kilauea in 1789 the ground shook and rocked so that persons could not stand.

The first eruption of the volcano Irasu, in Costa Rica (1783), was accompanied by violent earthquakes.[1] The smoke and flames which are said to have issued from the side òf Mount Fojo at the time of the Lisbon earthquake are regarded by some as having been volcanic. Others thought that the phenomena, rather than being on the side of Fojo, which showed no traces of volcanic action, had taken place in the ocean.

At the time of the great earthquake at Concepcion (1835), whilst the waves were coming in, two great submarine eruptions were observed. One, behind the Isle of Quiriquina, appeared like a column of smoke. The other, in the bay of San Vicente, appeared to form a whirlpool. The sea-water became black, and had a sulphurous smell, there being a vast eruption of gas in bubbles. Many fish were killed.[2]

With this same earthquake, near to Juan Fernandez, about one mile from the shore, the sea appeared to boil, and a high column of smoke was thrown into the air. At night flames were seen.

In 1861, when Mendoza was destroyed and 10,000 inhabitants killed, a volcano at the foot of which Mendoza is situated burst into eruption.

The earthquake of 1822 at Valdivia was accompanied by eruptions of the neighbouring mountains, which only lasted a few minutes.

At the time of the Leghorn shocks (January 16–27, 1742) some fishermen observed a part of the sea to rage

[1] *Jour. Royal Geog. Soc.* vol. vi.   [2] *Ibid.* vol. vi.

violently, to raise itself to a great height, and then rush landwards.[1]

In 1797, when Riobamba was destroyed, the neighbouring volcanoes were not affected, but Mount Pasto, 120 miles distant, suddenly ceased to throw out its usual column of water.

On the night of December 10, 1874, a strong shock was felt in New England, whilst at 4.45 A.M. on December 11 a shock was felt in the Pic du Midi, in the Pyrenees. In the middle of December there were volcanic outbursts in Iceland.[2]

It is possible that these occurrences might be the results of some widespread disturbance beneath the crust of the earth, or perhaps even of widely extended earth pulsations. The probability, however, is that these coincidences are accidental. When we remember that in a small area like the northern half of Japan alone there are periods when there are at least two shocks per day on the average, it is impossible for these coincidences not to exist. Less frequently coincidences between the larger disturbances must occur. Over and above these accidental coincidences, it would appear that in the world's history periods have occurred when earthquakes were unusually frequent, and at such times distant countries have suffered simultaneously. This approximate coincidence in period, which has been referred to when speaking of the distribution of destructive earthquakes in historical time, does not imply an exact synchronism in the single shocks.

Small earthquakes, or, more properly speaking, local tremblings, are a necessary accompaniment of almost all volcanic eruptions. Tremors of this description are seldom, however, felt beyond the crater, or at the most upon the flanks of the mountain where the eruption is going on,

[1] *Phil. Trans.* vol. xlii.   [2] *Am. Jour. Sci.* vol. x. p. 191.

They are due to the explosive action of steam bursting through the molten lava.

Volcanic eruption succeeding earthquakes.—Sometimes has happened that an earthquake, or a series of earthquakes, have terminated with the formation of volcanic vents.

As an example of a volcanic outburst terminating a seismic disturbance, may be mentioned the appearance of a new volcano in the centre of Lake Ilopango, as a sequel to the shocks which had disturbed that neighbourhood in 1879.[1]

In 1750 there were continuous shakings lasting over three months at Manilla. These terminated with an eruption of a small island in the middle of a neighbouring lake. Three days after the commencement of this eruption, four other small islands rose in the same lake.[2]

Antonio d'Ulloa, when speaking of the Andes, remarks that after a volcanic eruption the shocks cease.[3]

Conclusion.—Looking at this question generally, insomuch as the greatest number of volcanic eruptions appear, according to Fuchs, to have taken place in summer, whilst the greatest number of its earthquakes have apparently taken place in winter, it would seem that the two phenomena are without any direct connection, unless it be that both are different effects of a common cause.

Regarded in this manner, an earthquake may be looked upon as an uncompleted effort to establish a volcano. To use the words of Mallet, ' The forces of explosion and impulse are the same in both ; they differ only in degree

---

[1] 'Earthquakes of San Salvador, December 21–30, 1879.' *Am. Jour. Sci.* vol. xix. p. 415.

[2] *Gent. Mag.* 1757, p. 323.   [3] *Phil. Trans.* vol. li., 1760.

of energy, or on the varying sorts and degrees of resistance opposed to them.'[1]

Although we have many examples of earthquakes having occurred without volcanic eruptions, and, on the other hand, of volcanic eruptions without earthquakes, volcanoes may still be regarded as 'safety-valves of the earth's crust,' which, by giving relief to internal stresses, guard us against the effects of earthquakes.

That many earthquakes are felt at Copiapo is attributed to the fact that in the neighbouring mountains there are no volcanic vents.

We must not, however, overrate the protective influence of volcanoes. In the Sandwich Islands we see the columns of liquid lava in neighbouring mountains standing at different heights, indicating a want of subterranean connection between these vents. In consequence of this it would seem that enormous pressures might be generated in the neighbourhood of one of these mountains without finding relief at the other. When we have conditions like these, it would seem that the eruption of a volcano may have little or no influence in protecting neighbouring districts.

This may possibly be the explanation of the fact that in 1835 Concepcion was destroyed, notwithstanding there being an unusual activity in the volcanic vents of the neighbouring mountains.

[1] Mallet, *Report to Brit. Ass.*, 1858, p. 67.

CHAPTER XVII.

THE CAUSE OF EARTHQUAKES.

As the results of modern inquiries respecting the
cause of earthquakes, we see many investigators chiefly
attributing these phenomena to special causes. A few
attribute them to several causes. It seems to us that
they might be attributed to very many causes which often
act in a complex manner. The primary causes are telluric
heats, solar heat, and variations in gravitating influences.
These may be the principal, and sometimes the immediate, cause of an earthquake. The secondary causes are
those dependent upon the primary causes, such as expansions and contractions of the earth's crust, variations
in temperature, barometrical pressure, rain, wind, the
attractive influences of the sun and moon in producing
tides in the ocean or the earth's crust, variations in the
distribution of stress upon the earth's surface caused by
processes of degradation, the alterations in the position
of isogeothermal surfaces, &c.

The part which may be played by these various causes

in the production of oscillations, pulsations, and tremors will be referred to.

Earthquakes consequent on faulting.—In the chapter on Earth Oscillations, the causes producing the phenomena of elevation and depression are briefly indicated.

By the variations in stress accompanying elevations and depressions, cracks are produced. Inasmuch as compression would crush the rocks constituting the earth's crust, we must conclude with Captain Dutton that these cracks are formed by tension. By elevation, the upper rigid crust of the earth is stretched, and fissures are produced. The sudden formation of these fissures or faults gives rise to earthquakes, and perhaps also to volcanic vents. That earthquake and volcanic regions are situated on areas where there is evidence of rapid elevation is strikingly illustrated round the shores of the Pacific.

Lasaulx considered that the earthquake of Herzogenrath was more or less intimately connected with the great mountain fissure—the *Feldbiss*—which crosses the coal region of the Wurm.[1] The sudden elevation or sinking of large areas at the time of an earthquake may be a consequence of these dislocations.

It has already been pointed out that the earthquake region of Japan is the one where we have evidence of recent and rapid elevation. That certain earthquakes of this region may possibly be the result of faulting we have the evidence of our senses and of our instruments. The sudden blows and jolts which are sometimes felt are indicative of the sliding of one mass of rock across another.

Should the ground be simply torn asunder, this tearing would give rise to a series of waves of distortion, vibrating in directions parallel to the plane of the fissure. Supposing this motion to be propagated to a number of

[1] Von Lasaulx, *Earthquakes of Herzogenrath.*

surrounding stations, it would be recorded at each of these as having the same direction. To those situated on a line forming a continuation of the strike of the fissure, the vibrations would advance so to speak *end on*, whilst to those stations lying in a line perpendicular to the strike of the fissure, the motion would advance *broadside on.*

Motions like these latter have been recorded in Tokio, where earthquakes which from time observations were known to have come from the faulted and rising region to the south have been registered as a series of east and west motions, or vibrations transverse to this line of propagation.

It must, however, be here mentioned that the registration of only transverse motion may possibly be due to the extinction of normal motion, although this is not generally regarded as probable.

It would therefore appear that certain earthquakes and faults are closely related phenomena, the former being an immediate effect of the latter. Faults are due to earth oscillations, and to a variety of causes producing disturbances in the equilibrium of the earth's crust ; the principal cause of all these phenomena being alterations in the distribution of heat, and the attractive force of gravity.

Earthquakes consequent on the explosion of steam.— Humboldt regarded volcanoes and earthquakes as the results of a common cause, which he formulated as ' the reaction of the fiery interior of the earth upon its rigid crust.' Certain investigators, who have endeavoured to reduce Humboldt's explanation to definite limits, have suggested that earthquakes may be due to sudden outbursts of steam beneath the crust of the earth, and its final escape through cracks and fissures.

Admitting that steam may accumulate by separating

out from the cooling interior of our globe, its sudden explosion might be brought about by its own expansive force, or by the movements in the bubbling mass from which it originated.

Others, however, rather than regard the steam as being a primeval constituent of the earth's interior, imagine it arises from the gradual percolation of water from the surface of the earth down to volcanic foci, into which it is admitted against opposing pressures, by virtue of capillary action.

Mallet, in his account of the Neapolitan earthquake, shows that the whole of the observed phenomena can be accounted for by the admission of steam into a fissure, which by the expansive force exerted on its walls was rent open. Just as at the Geysers we hear the thud and feel the trembling produced by the sudden evolution and condensation of steam, so may steam by its sudden evolution and condensation in the ground beneath us give rise to a series of shocks of varying intensity, accompanied by intermediate vibratory motions—that is to say, a motion which, as judgèd of by our feelings, is not unlike many earthquakes. Often it may happen that the result of the explosion may be the production of a fault, or at least a fissure ; and thus in the resulting movements we may have a variety of vibrations, some being those of compression and distortion, produced by the blow of the explosion, and others being those of distortion alone, produced by the shearing action which may have taken place by the opening of the fault. Sometimes one set of these vibrations may be prominent, and sometimes the other. Thus, when we say that an earthquake has shown evidence by the nature of its vibrations that it was produced by a fault, this by no means precludes the possibility that an explosion of steam may also have been connected with the production of the disturbance. Mallet threw out the suggestion that the opening of fissures beneath the ocean might admit water to volcanic foci. During the time that the water was in the spheroidal state, the preliminary tremors, so common to many earthquakes, would be produced. These would be followed by the explosion, or series of explosions, constituting the shock or shocks of the earthquakes.

The chief reasons for believing that the earthquakes of North-Eastern Japan are partly due to explosive efforts are :—

1. That the greater number of disturbances, perhaps ninety per cent., originate beneath the sea, where we may imagine that the ground, under the superincumbent hydrostatic pressure, is continuously being saturated with moisture.

2. Many of the diagrams show that the prominent vibrations, of which there are usually from one to three, in a given disturbance have the same character as those produced by an explosive like dynamite, the greatest and probably the most rapid motions being inwards towards the origin.

It may here be remarked that a very large proportion of the destructive earthquakes of the world have originated beneath the sea, as has often been testified by the succeeding sea waves. Also, it must be observed, that earthquake countries, like volcanic countries, are chiefly those which have a coast line sloping at a steep angle beneath the sea—that is to say, earthquakes are frequent along coasts bordered by deep water.

The earthquakes which occur at volcanic foci constitute another class of disturbances which may be accredited to the explosive efforts of steam.

Earthquakes due to volcanic evisceration.—By the

ejection of ashes and lava from volcanic vents, there is
an extensive evisceration of the neighbouring ground.
When we look at a volcano like Fujiyama, 13,000 feet
in height, and at least fifty miles in circumference, and
remember that the mass of cinders and slag of which it
is composed came from beneath the area on which it
rests, the point to be wondered at is, that earthquakes,
consequent on the collapse of subterranean hollows, are
not more frequent than they are. At the time of a single
eruption of a volcano, the quantity of lava ejected amounts
to many thousand millions of cubic feet. In 1783 the
quantity of lava ejected from Skaptas Joknee, in Iceland,
was estimated as surpassing 'in magnitude the bulk of
Mont Blanc.' [1] Admitting that hollow spaces are the
results of these eruptions, and that in consequence of
this evisceration the ground is rendered unstable, the
instability being increased by the additional load placed
above the eviscerated area, it would seem that from time
to time earthquakes are inevitable.

Facts, however, teach us that volcanoes act as safety
valves, and that, as a rule, at or shortly after an eruption,
earthquakes cease. The relationship of earthquakes to
volcanic eruptions would therefore indicate, notwithstanding the arguments put forward to show that an area
loaded by a volcano has in consequence of the evisceration
and the load a quaquaversal dip, that evisceration does
not take place beneath volcanoes as is usually supposed,
and we may conclude that it is but few earthquakes
which have an origin due to these causes.

*Earthquakes and evisceration by chemical degrada-
tion.*—A powerful agent, which tends to the formation
of subterranean hollows, is chemical degradation. The
effects of this have been often measured by quantitative

[1] Lyell, *Principles*, vol. ii. p. 51.

analysis of the solid materials which are daily carried away by many of our springs. In limestone districts this is very great. Prof. Ramsay estimates that the mineral matter discharged annually by the hot springs of Bath is equivalent in bulk to a column 140 feet in height and 9 feet in diameter. At San Filippo, in Tuscany, the solid matter discharged from the springs has formed a hill a mile and a quarter long, a third of a mile broad, and 250 feet in thickness.[1] Many other examples of subterranean chemical degradation will be found in text-books of geology.

By this chemical action large cavernous hollows are produced. Beneath a volcano it is probable that liquid material immediately takes the place of that which is ejected, and that hollows are not formed as in the case of chemical degradation. If a cavern becomes too large, it eventually collapses.

Of the falling in of large excavations we have examples in large mines. As a consequence, not only is a trembling produced, but also a noise, which is so like that produced by certain earthquakes that the South American miners have but one word, ' bramido,' to express both.[2]

Boussingault, who was an advocate for the theory that many earthquakes are produced by the sinking of the ground, calls attention to the fact that we have evidences of the subsidence of great mountains, like the Andes, the districts around which are so well known for their earthquakes. Capac Urcu is one of these mountains which legends tell us has decreased in height.

The variation in the height of mountains is a subject which deserves attention. That mountains may possibly be hollow, we have the remarkable results attained by

[1] Lyell, *Principles*, vol. i. p. 402.　　　[2] Fuchs, p. 464.

Captain Herschel, who found that the attractive force of gravity in the neighbourhood of the Himalayas was not so great as it ought to have been had these mountains been solid. The Rev. O. Fisher gives another explanation of this phenomenon. Palmieri considers that the terrible earthquake which devastated Casamicciola (1881) was due to the hot springs having gradually eaten out cavernous spaces beneath the town. The extremely local character of this shock was certainly favourable to such a view.

The earthquake which, in 1840, caused Mount Cernans, in the Jura, to fall, is also attributed to the solvent action of waters in undermining its foundations. This undermining action was in great measure probably due to a large spring, which, twenty-five years previously, had disappeared, and which subsequently may possibly have been slowly disintegrating the foundations of the mountain. Earthquakes of this order would be principally confined to districts where there are rocks which are more or less soluble, as, for instance, rock salt, gypsum, and limestone.

Earthquakes and the attractive influences of the heavenly bodies.—The most important attractions exercised upon our planet are those due to the sun and moon. To these influences we owe the tides in our ocean, and possibly elastic tides in the earth's crust. Some theorists would also insist upon liquid tides in the fluid interior of our earth. The nature of the earth's interior is, however, a question on which there is a diversity of opinion.

One doctrine, which, until recent years, received much support, was that the interior of the earth was a reservoir of molten matter contained within a thin crust. Hopkins showed that the least possible thickness of such a crust must be from 800 or 1,000 miles, otherwise the motions of precession and nutation would be subject to interference.

M. Delauney objected to the views of Hopkins, on the supposition that the fluid interior of the earth had a certain viscosity.

Sir William Thomson arrives at the conclusion that the earth on the whole must be more rigid than a continuous solid globe of glass. Mr. George H. Darwin's investigations on the bodily tides of viscous or semi-elastic spheroids tend to strengthen the arguments of Sir William Thomson.

Some philosophers hold the view that the central portion of the earth, although intensely hot, is solid by pressure, whilst the outer crust is solid by cooling. Between the two there is a shell of liquid or viscous molten matter.

Another argument is, that although the interior of the globe may be solid, it is only retained in that condition by an immense pressure, on the relief of which it is liquefied—it is potentially liquid.

As these views, and the arguments for and against them, are to be found in all modern text-books of geology, we will at once proceed to consider the effect of solar and lunar attractive influences in producing earthquakes upon a globe which is either solid, partially solid, or which has an interior wholly liquid.

Effect of the attractive influences of the sun and moon. —In 1854 M. F. Zantedeschi put forward the view —that it is probable there is a continual tendency of the earth to protuberance in the direction of the radii vectores of the two luminaries which attract it. In consequence of these protuberances, pendulums ought at one time to swing more slowly than at others. Zantedeschi remarks that the periods of earthquakes appear to confirm such a view, insomuch as they occur more often at the syzygies, or epoch of the spring tides, than at neap tides—an observation found in the works of Georges Baglivi (1703) and Joseph Toaldo (1770).[1]

Prof. Perrey, of Dijon, who did so much for seismology, held the view that the preponderance in the number of earthquakes felt at particular seasons was possibly due to the attractive influence of the sun and moon producing a tide in the fluid interior of the earth, which, acting on the solid crust, produced fractures.

Rudolf Falb, whose writings have of late years attracted considerable attention, brings forward views which may be regarded as amplifications of those suggested by Perrey.

According to Falb, the inner portion of the earth must be regarded as fluid. In the crust above this fluid reservoir are cracks and channels, into which, by the attraction of the moon and sun, the fluid is drawn. On entering these cracks cooling takes place, together with explosions of gas and subterranean volcanic disturbances. The attractions producing the internal tides required by Falb are chiefly dependent upon the following factors :—

1. The nearness and distance of the sun from the earth (January 1 and July 1).

2. The position of the moon with regard to the earth, which in every twenty-seven days is once near and once distant.

3. The phases of the moon—whether full or new moon (syzygies), or whether first or last quarter (quadratures).

4. The equinoxes, the position of the sun in the equator, and the relative position of the earth.

5. The position of the moon relative to the equator.

[1] *Comptes Rendus,* August 1854.

6. The concurrence of the 'centrifugal force' of the earth with the last quarter of the moon.

7. The entrance of the moon on the ecliptic—the so-called nodes.

Assuming that earthquakes are wholly consequent on these attractions, it at once becomes possible to predict their occurrence. This Falb does, and when his predictions have been fulfilled he has certainly gained notoriety.

He commenced by the predictions of great storms. In 1873 he predicted the destructive earthquake of Belluno, which earned for himself a eulogistic poem, which he has republished in his 'Gedanken und Studien über Vulcanismus.' After this, in 1874, he predicted the eruption of Etna. He also explained why, in B.C. 4000, there should have been a great flood, and for A.D. 6400 he predicts a repetition of such an occurrence.

When we approach the question of the extent to which the attraction of the sun and moon may influence the production of earthquakes, a question which we have to answer is, whether it is likely that the attractive power of the moon is so great that it could draw up the crust of the earth beyond its elastic limits. We know what it can do with water. It can lift up a hemispherical shell 8,000 miles in diameter about two or three feet higher at its crown than it lifts the earth. Even supposing the solid crust to be lifted 100 times the apparent rise of the tide, is it likely that a hemispherical arch 8,000 miles in diameter when it is raised 200 feet at its crown could by any possibility suffer fracture ? If an arch is 12,000 miles in length, all that we here ask is, whether the materials which compose the arch are sufficiently elastic to allow themselves to be so far stretched that the crown may be raised 200 feet. The result which we should arrive at is apparently so obvious that actual calculation seems hardly

necessary. If we regard the earth as being solid, the
question resolves itself into the inquiry as to whether a
column of rock, which is equal in length to the diameter
of the earth, or about 8,000 miles, can be elongated 200
feet without a fracture. This is equivalent to asking
whether a piece of rock one yard in length can be
stretched one seventy thousandth of a foot. Considering
that this is a quantity which is scarcely appreciable under
the most powerful of our microscopes, we must also
regard this as a question which it is hardly necessary to
enter into calculations about before giving it an answer.
To vary the method of treating such a question, may we
not ask what is the utmost limit to which it would be
possible to raise up or stretch the crust of the earth
without danger of a fracture? Thus, for instance, to
what extent might a column of rock be elongated without
danger of its being broken? From what we know of the
tenacity of materials like brick and their moduli of elasticity, it would seem possible to stretch a bar of rock
8,000 miles in length for approximately half a mile
before expecting it to break. As to whether there is a
wave, the height of which is equal to half this quantity,
running round our earth as successive portions of its
surface pass beneath the attracting influences of the sun
and moon, is a phenomenon which, if it exists, would
probably long ago have met with a practical demonstration.

The deformation which a solid globe or spherical shell
would experience under the attractive influences of the
sun and moon has been investigated by Lamé, Thomson,
Darwin, and other physicists and mathematicians.

A conclusion that we are led to as one result of these
valuable investigations is, that if the interior of the earth
be fluid, and covered with a thin shell, then enormous

elastic tides must be produced. A consequent phenomenon, dependent on the existence of these tides, would
be a marked regularity in the occurrence of earthquakes.
As this marked regularity does not exist, we must
conclude that earthquakes are not due to the attractive
influences of the sun and moon acting upon the thin
crust of the earth covering a fluid interior. The periodicity of earthquakes corroborates the conclusions of Sir
William Thomson, who remarks that if the earth were
not extremely rigid the enormous elastic tides which
must result would be sufficient to lift the waters of the
ocean up and down so that the oceanic tide would be
obliterated.

Assuming that the earth has the rigidity assigned to it
by mathematical and physical investigators, we nevertheless have travelling round our earth, following the attractions of the moon and sun, a tidal stress. This stress, imposed upon an area in a critical state, may cause it to give
way, and thus be the origin of an earthquake. Earthquakes ought therefore to be more numerous when these
stresses are the greatest.

The periods of maximum stress or greatest pull exerted by the moon and sun will occur when these bodies
are nearest to our planet—that is, in perigee and perihelion, and again when they are acting in conjunction
or at the syzygies. That earthquakes are *slightly* more
numerous at these particular periods than at others
is a strong reason for believing that the attractions of
the moon and sun enter into the list of causes producing
these phenomena.

Had there been a strongly marked distinction in
the number of earthquakes occurring at these particular
seasons as compared with others, we might have attributed
earthquakes to the existence of elastic tides of a sensible

U

magnitude. As the facts stand, it appears that the maximum pulls exerted by the moon and sun are only sufficient to cause a slight preponderance in the number of earthquakes felt at particular seasons, and therefore that these pulls only result in earthquakes when the distorting effort has been exerted on an area which, by volcanic evisceration, the pressure of included gases, and other causes, is on the verge of yielding.

Earthquakes and the tides.—If we assume that earthquakes are in many cases due to the overloading of an area and its consequent fracture, such loading may occur by the rising of the tide. A belief that the earthquakes of Japan were attributable to the tides may be found in the diary of Richard Cocks under the date November 7, 1618, who remarks:—

'And, as we retorned, about ten aclock, hapned a greate earthquake, which caused many people to run out of their howses. And about the lyke hower the night following hapned an other, this countrey being much subject to them. And that which is comunely markd, they allwais hapen at a hie water (or full sea); so it is thought it chauseth per reason is much wind blowen into hollow caves under ground at a loe water, and the sea flowing in after, and stoping the passage out, causeth these earthquakes, to fynd passage or vent for the wind shut up.'[1]

Although we may not acquiesce in Cock's views respecting the imprisoned wind, it would seem that a comparison of the occurrence of earthquakes and the state of the tide would be a legitimate research. Inasmuch as the stresses which are brought to bear upon an area by the rising of the tide are so very much greater than those due to barometrical changes, it is not unlikely

[1] *Nature*, April 26, 1883.

that a marked connection would be found. But it must be remembered that because researches, so far as they have gone, tend to show that earth movements are more frequent when an area is relieved of a load, it is not unlikely that the greatest number of earthquakes may be found to occur at low water. Prof. W. S. Chaplin attempted to make this investigation in Japan, but not being able to obtain the necessary information respecting the tides, was compelled to relinquish this interesting work.

Every foot of rise in a tide is equivalent to a load being placed on the area over which the tide takes place of sixty-two pounds to the square foot. This load is not evenly distributed, but stops abruptly at a coast line. Lastly, it may be observed that many coast lines are not simultaneously subjected to stresses consequent upon this load. Japan, for instance, may be regarded as an arch placed horizontally. The area near the crown of this arch is loaded by the tidal wave crossing the Pacific before the areas near the abutment, and farther there is a horizontal pressure at the crown which, if Japan were like a raft, would tend, as the tide advanced, to straighten its bow-like form, but as the wave passed its abutments to increase its curvature.

Prof. G. Darwin has calculated the amount of rise and fall of a shore line due to tidal loads (see p. 336, ' Earth Pulsations '). The result of these calculations apparently indicates that these loads may have a considerable influence upon the stability of an area in a more or less critical condition.

Mr. J. Carruthers suggests that tidal action may hold a general but indirect relationship to volcanic and seismic action by the retardation it causes on the earth's rotation. By this retardation the polar axis tends to lengthen, and

tensile stresses are induced, resulting in fracture. The
fluid interior of the earth, being no longer restrained,
would move polewards, and, leaving equatorial portions
unsupported, this would gradually collapse. The primary
fractures would be north and south, while the secondary
fractures would be east and west.[1]

That the rise of the tide is accompanied by a greater
percolation of water to volcanic foci, which, in consequence, assume a greater state of activity, is a theory
which was advanced many years ago. To determine
how far tides may directly be connected with earthquakes,
the necessary records have yet to be examined.

Variations in atmospheric pressure.—When we
consider the immense load which, by a sudden rise of
the barometer, is placed upon the area over which this
rise takes place, it is not difficult to imagine that this
rise may occasionally be the final cause which makes the
crust of the earth to give way. A barometric rise of
an inch is equivalent to a load of about seventy-two
pounds being put upon every square foot of area over
which this rise takes place. On the other hand, a fall
in the barometric column indicates that a load has been
removed, and whatever elastic effort may be exerted by
subterranean forces in endeavouring to escape, being
met by less resistance, they may burst these bonds, and
an earthquake will result. For reasons such as these
the final cause of earthquakes has often been attributed
to variations in atmospheric pressure. In Japan there
are practically as many earthquakes with a high barometer
as with a low one.

The extent to which barometric fluctuations have
acted as final causes in the production of earthquakes
may be judged of by a comparison of the times of baro-

[1] *Phil. Soc.*, Wellington, New Zealand, 1875.

metric variation and the times at which earthquakes have occurred.

Three important laws of barometric variation are the following :—

1. In the world generally the average barometric pressure is highest in winter. (Exceptions occur near Iceland and in the North Pacific.)

2. The summer and winter monthly mean barometer differs least near the equator and over the great oceans. They differ most over the great continents and generally with increasing latitude.

3. The greatest number of barometrical fluctuations usually take place in winter.

Inasmuch as there are generally more earthquakes in winter than in summer, the first of these laws would indicate that this might be due to the greater load which acts upon the crust of the earth at that season. The second law would indicate that the distinction between the winter and summer earthquakes ought to be most marked in high latitudes, which, if we refer to the table on p. 257, we observe to be borne out by the results of observation. The countries where there are as many earthquakes in winter as in summer are chiefly those in low latitudes. The number of these countries from which we have records are, however, few.

Facts opposed to the idea that earthquakes may be caused by an increase of barometric pressure are the results of observations like those of Schmidt and Rossi, which show that earthquakes chiefly occur with a low barometer.

Assuming that these latter observations will be found by future investigators to be generally true, we must conclude that the relief of atmospheric pressure has an influence upon the occurrence of earthquakes. Such a

conclusion would partially accord with the third barometrical law, or the fact that there are more occasions on
which we get a low barometer during the winter months.

Other writers who have examined this question are
Volger, Kluge, Andrès, and Poly. The latter investigator
sought a connection between earthquakes and revolving
storms, in the centres of which there is usually an abnormal decrease of atmospheric pressure. If an area over
which such a sudden change in pressure took place was
in a critical state, it is not difficult to see that storms
such as Poly refers to might sometimes be accompanied
by earthquakes.

Fluctuations in temperature.—Inasmuch as fluctuations in temperature are governed by the sun, it may at
once be said that there is a connection between earthquakes and readings of the thermometer. Certainly
earthquakes occur mostly during the cold months or
in winter. Similarly, as changes in temperature are so
closely connected with barometric fluctuations, and these
are said to have a direct influence upon the yielding of
the earth's crust, seismic phenomena are indirectly linked
to fluctuations in temperature. A rise in temperature is
usually accompanied by a fall in the barometer, and this
in turn may be a condition favourable for the occurrence
of an earthquake.

If we regard solar heat as an agent causing expansions
or contractions in the earth's crust, then fluctuations in
temperature become an immediate cause of earthquakes.
The probability, however, is that solar heat has little or
no connection with the final cause producing earthquakes,
although at the same time coincidences between the
occurrence of earthquakes and unusual fluctuations in
temperature may from time to time be observed.

Winds and earthquakes.—Although it may be

admitted that high winds exert enormous pressures upon mountain ranges, and might occasionally give rise to stresses causing rocky masses in unstable equilibrium to give way, the coincidences which have been established between the occurrence of storms and earthquakes can usually only be regarded as occurrences which have synchronised by chance.

Storms are usually accompanied with a barometric depression, and the relation of diminutions in atmospheric pressure to earthquakes has been discussed.

Rain and earthquakes.—It has already been shown that earthquakes have occasionally been found to coincide with rain and rainy seasons. Whether the saturation of the ground with moisture or the percolation of the same to volcanic foci may be a direct effect producing earthquakes it is difficult to say. The probability, however, is that, rain being dependent on phenomena like changes in temperature, barometric fluctuations, and winds, we must regard it and the earthquakes which happen to coincide with these precipitations of moisture as congruent effects of more general causes.

Conclusion.—Although it would be an easy matter to discuss the relationship of earthquakes and other phenomena, we must conclude that the primary cause of earthquakes is endogenous to our earth, and that exogenous phenomena, like the attraction of the sun and moon and barometric fluctuations, play but a small part in the actual production of these phenomena, their greatest effect being to cause a slight preponderance in the number of earthquakes at particular seasons. They may, therefore, sometimes be regarded as final causes. The majority of earthquakes are due to explosive efforts at volcanic foci. The greater number of these explosions take place beneath the sea, and are probably due to the admission of water through fissures to the heated rocks beneath. A smaller number of earthquakes originate at actual volcanoes. Some earthquakes are produced by the sudden fracture of rocky strata or the production of faults. This may be attributable to stresses brought about by elevatory pressure. Lastly, we have earthquakes due to the collapse of underground excavations.

CHAPTER XVIII.

PREDICTION OF EARTHQUAKES.

General nature of predictions—Prediction by the observation of unusual phenomena (alteration in the appearance and taste of springs;
underground noises; preliminary tremors; earthquake prophets
—warnings furnished by animals, &c.)—Earthquake warning.

General nature of predictions.—Ever since seismology
has been studied, one of the chief aims of its students has
been to discover some means which would enable them to
foretell the coming of an earthquake, and the attempts
which have been made by workers in various countries to
correlate these occurrences with other well-marked phenomena may be regarded as attempts in this direction.

Ability to herald the approach of these calamities
would unquestionably be an inestimable boon to all who
dwell in earthquake-shaken countries, and the attempts
which have been made both here and in other places are
extremely praiseworthy. In almost all countries where
earthquakes are of common occurrence these movements
of the earth have been more or less connected with
certain phenomena which, in the popular mind, are
supposed to be associated with an approach of an
earthquake.

If predictions were given in general terms, and they
only referred to time, inasmuch as on the average there
are in the world several shakings per day, we should

always find that predictions were verified. We might even go further and predict that on certain days earthquakes would occur in certain countries, and still find that in many instances our supposed power of foresight had not deceived us. Thus, for instance, in Japan, where on the average there are probably one or two shakings every day, if prognostications were never correct there would be a violation of the laws of chance.

What is required from those who undertake to forewarn us of an earthquake is an indication not only of the time at which the disturbance will happen, but also an indication of the area in which it is to occur. Those who dwell in an area where there are certain well-defined periods during which seismic activity is at a maximum—if ten or fourteen days should have passed without a shock—might, in many instances, find that a prophecy that there would be an earthquake within the next few days would prove itself correct. Also, if a severe shock had taken place, a prophecy that there would be a second or third smaller disturbance within a short period would also meet with verification.

Certain persons with whom I am intimate appear to have persuaded themselves that they can foretell the coming of an earthquake by the sultry state of the atmosphere or a certain oppressiveness they feel, and an instinctive feeling arises that an earthquake is at hand.

It is said that a few minutes before many of the shocks which shook New England between 1827 and 1847 people could foretell the coming disturbance by an alteration in their stomach.[1] No doubt many who dwell in earthquake countries, and have been alarmed by earthquakes, are at times subject to nervous expectancy.

The author has had such sensations himself, due, per-

[1] *Phil. Trans.*, vol. xlii.

haps, to a knowledge that it was the earthquake season, that there had been no disturbance for some weeks, and a consequent increasing state of nervous presentments. In consequence of this, not only has he carefully prepared his instruments for the coming shock, but he has written and telegraphed to friends to do the same.

Sometimes these guesses have proved correct. One remarkable instance was a few hours prior to the severe shock of February 22, 1880, when he communicated with his friends in Yokohama and asked them to see that their instruments were in good order. Oftener, however, his prognostications have been incorrect. The point in connection with this subject which he wishes to be remarked is, that the instances where earthquakes occurred shortly after the receipt of his letters are carefully remembered, and often mentioned, but the instances in which earthquakes did not occur appear to be entirely forgotten. He is led to mention these facts because they appear to be an experimental proof of what has taken place in bygone times, and what still takes place, especially amongst savages—namely, that the record of that which is remarkable survives, whilst that which is of every-day occurrence quickly dies. Had the records of all prognostications been preserved, the probability is that we should find that they had, in the majority of cases, been incorrect, whilst it would have been but in very few instances they had been fulfilled.

Prediction by the observation of natural phenomena.--The above remarks may perhaps help us to understand the prognostications of the ancient philosophers about which Professor Antonio Favaro, of Padua, has written.[1] Cicero in the ' De Divinatione,' speaking on this subject, says that ' God has not predicted so much

[1] M. S. di Rossi, *Earthquakes of Casamicciola.*

as the divine intelligence of man.'—' Non Deus prævidet
tantum, sed et divini ingenii viri.' Favaro regards these
predictions, however, as the result of observations of nature
which show it is possible that indications of coming earthquakes had been announced by variations in the gas given
out from subterranean sources, the change in colour,
taste, level, temperature of the water in springs, &c.

In 1843 a bishop of Ischia forewarned his people of a
coming earthquake, and thus was instrumental in the
saving of many lives. Naturally, in an age of superstition,
the bishop would be regarded as a prophet, but Favaro
considers that the prognostication was probably due to a
knowledge of premonitory signs as exhibited in changes
in the characters of mineral waters.

The shock of 1851, at Melpi, was in this way predicted
by the Capuchin fathers, who observed that a lake near
their door became frothy and turbulent.

Underground noises have led persons to the belief
that an earthquake was at hand. It was in this way that
Viduari, a prisoner at Lima, predicted the destruction of
that city.

Before the earthquake of 1868, so severely felt at
Iquique, the inhabitants were terrified by loud subterranean noises.

That underground noises have preceded earthquakes
by considerable intervals appears to be a fact, but, at the
same time, it must be remembered that similar noises
have often occurred without an earthquake having taken
place.

Farmers predicted the earthquake of St. Remo, in
1831, by underground noises.

On the day before the earthquake which, in 1873,
shook Mount Baldo, the inhabitants of Puos, a village
north of Lake Santa Croce, heard underground noises.

Before the earthquakes which, in 1783, shook Calabria and Sicily, fish are said to have appeared in great numbers on the coast of Sicily, and the whirlpool of Charybdis assumed an unusual excited state.

It is said that Pherecydes predicted the earthquakes of Lacedemon and Helmont, by the taste of the water in the very deep well at the castle of Lovain.[1]

The writer of an article on the Lisbon earthquake says that 'after the 24th I felt apprehensive, as I observed the same prognostics as on the afternoon of October 31, that is, the weather was severe, the wind northerly, a fog came from the sea, the water in a fountain ran of a yellow clay colour, and' he adds, 'from midnight to the morning of the 25th I felt five shocks.'[2]

At the present time Rudolf Falb, following a theory based upon the attractive influences of the sun and moon, tells us the time at which we are to expect earthquakes.

That occasionally there are signs attendant on earthquakes, although we cannot give them a physical explanation, we cannot doubt. Also we know that in certain areas earthquakes are more likely to occur at one season than at another. Should earthquakes be foretold with the assistance of knowledge of this description, the predictions at once become the result of the application of certain natural laws, and are not to be regarded as predictions in the popularly accepted sense of that term, any more than the arrival of a friend is predicted by the previous receipt of a telegram announcing his coming.

Rather than accredit the ancients and those of more modern times who, in consequence of their feelings, have recorded the coming of an earthquake, with a knowledge of premonitory signs, we might in many instances regard

[1] *Phil. Trans.*, vol. xviii. 1683-5.   [2] *Ibid.* vol. xlix.

the records of those prognostications as the survival of
accidental guesses, and, as such, examples of the survival
of the useless.

The effect of accidental occurrences of this description
upon an uneducated mind, in engendering superstition, is
a subject which has often been dwelt upon, and the difficulty of eradicating the same—as may be judged of by the
following accident which came under the observation of
Mr. T. B. Lloyd and the author, in 1873, when travelling
in Newfoundland—will be easily appreciated.

At the time to which I refer, my companion was
bringing a canoe down the rapids below the Grand Pond
in a country which is practically uninhabited, and where
an Indian trapper would perhaps be the only person met
with, and this not more than once a year. Whilst
shooting the rapids one of the Indians, Reuben Soulian,
shot at a deer passing up one bank of the river. That
the deer had been hit was testified by a trail of blood
which bespattered the rocks. Subsequently several more
shots were fired, and it was believed by all that the deer
was killed. Soulian quickly followed the animal to the
spot where it was supposed to have fallen. Some time
after he returned, having failed to find any trace of the
animal. He was greatly agitated, but eventually became
melancholy, saying that the sudden disappearance of the
animal was a sure sign that some of his relations had
suddenly died. About two hours afterwards Mr. Lloyd's
party met with a party of Indians coming up the river,
the first they had seen for four weeks, who told them
that Soulian's sister had just died on the coast.

In the northern part of South America certain shocks
are anticipated by preliminary vibrations which cause a
little bell attached to a T-shaped frame (cruz sonante) to
ring. There are, however, persons (trembloron) who are

supposed to be endowed with seismic foresight, whose verdicts are much relied upon.

In Caraccas it is said that nearly every street in the river suburb has an earthquake Cassandra or two. Some of these go so far not only as to predict the coming seisms, but also the vicissitudes of particular streets.[1] Earthquake prophets are, however, by no means confined to the new world, and many examples of them may be found in the histories of countries where earthquakes have been felt.

The story of the crazy lifeguardsman who prophesied an earthquake to take place in London on April 4, 1691, is an example. The Rev. Sig. Pasquel R. Perdini, writing on the earthquakes at Leghorn in 1742, says that 'a Milanese astrologer predicted this earthquake for January 27, by which " misfortune " the faith and credit given to the astrologer gained him more reverence and honour than the prophets and holy gospel.' Before the time at which he predicted a second shock, people removed away from Leghorn.

Warnings furnished by animals.—A study of the warnings furnished by animals is also interesting. Several of the natives in Caraccas possess oracular quadrupeds, such as dogs, cats, and jerboas, which anticipate coming dangers by their restlessness.

Before the catastrophe of 1812, at Caraccas, a Spanish stallion broke out from its stable and escaped to the highlands, which was regarded as the result of the prescience of a coming calamity. Before the disturbances of 1822 and 1835, which shook Chili, immense flocks of sea birds flew inland, as if they had been alarmed by the commencement of some suboceanic disturbance. Before this

---

[1] H. D. Warner, 'The City of Earthquakes,' *Atlantic Monthly*, March 1833.

last shock it is also related that all the dogs escaped from the city of Talcahuano.

Earthquake warning.—What has here been said respecting the prediction of earthquakes is necessarily imperfect—many of the signs which are popularly supposed to enable persons to foretell the coming of an earthquake having already been mentioned in previous chapters. That we shall yet be able to prepare ourselves against the coming of earthquakes, by the applications of laws governing these disturbances, is not an unreasonable hope.

With an electric circuit which is closed by a movement of the ground, we are already in a position to warn the dwellers in surrounding districts that a movement is approaching.

An earthquake which travelled at the rate of four seconds to the mile might, if it were allowed to close a circuit which fired a gun at a station fifteen miles distant, give the inhabitants at that place a minute's warning to leave their houses. The inhabitants of Australia and the western shores of the Pacific might, by telegraphic communication, receive eighteen to twenty-five hours' warning of the coming of destructive sea waves resulting from earthquakes in South America.

Although warnings like these might have their value, that which is chiefly required is to warn the dwellers at and near an earthquake centre of coming disturbances.

What the results of the observations on earth tremors will lead to is problematical.

Should microseismic observation enable us to say when and where the minute movements of the soil will reach a head, a valuable contribution to the insurance of human safety in earthquake regions will have been attained.

As to whether the movements of tromometers are destined to become barometric-like warnings of increased activity beneath the earth crust, or whether they are only due to vibrations of the earth crust produced by variations in atmospheric pressure, has yet to be investigated.

Other phenomena which may probably forewarn us of the coming of an earthquake are phenomena resultant on the stresses brought to bear upon the rocky crust previous to its fracture, or phenomena due to changes in the position and condition of heated materials beneath the earth's surface. Amongst these may be mentioned electrical disturbances, which appear to be so closely related to seismic phenomena.

At the time of earthquakes telegraph lines have been disturbed, but as to what may happen before an earthquake we have as yet but little information. The subject of earthquake warning is of importance to many countries, and is deserving of attention.

As our knowledge of earth movements, and their attendant phenomena, increases, there is but little doubt that laws will gradually be formulated, and in the future, as telluric disturbances increase, a large black ball gradually ascending a staff may warn the inhabitants on the land of a coming earthquake, with as much certainty as the ball upon a pole at many seaports warns the mariner of coming storms.

CHAPTER XIX.

EARTH TREMORS.

Artificially produced tremors—Observations of Kater, Denman, Airy, Palmer, Paul — Natural tremors — Observations of Zöllner, M. d'Abbadie, G. H. and H. Darwin—Experiments in Japan—With seismoscopes, microphones, pendulums — Work in Italy—Bertelli, Count Malvasia, M. S. di Rossi—Instruments employed in Italy—Tromometers, microseismographs, microphones—Results obtained in Italy—in Japan—Cause of microseismic motion.

DURING the past few years considerable attention has been drawn towards the study of small vibratory motions of the ground, which to the unaided senses are usually passed by without recognition. These motions are called *earth tremors*. Their discovery appears to have been due to accident, and not to the results of inductive reasoning. No sooner had philosophers contrived astronomical and other instruments for the purpose of making refined measurements and observations than they at once discovered that they had an enemy to contend against in the form of microscopic earthquakes.

Artificially produced tremors.—Artificial disturbances of this description exist in all our towns, and near a railway line they are perceptible with every passing train. Those who have used microscopes of high power will readily appreciate how small a disturbance of the ground is visible in the apparent movement of the object under examination.

Captain Kater found that he could not perform his pendulum experiments in London on account of the vibrations produced by the rolling of carriages. Captain Denman, who made some observations on artificially produced tremors, found that a goods train produced an effect 1,100 feet distant in marshy ground over sandstone. Vertically, however, above a tunnel through the sandstone, the effects only extended 100 feet.

A remarkable example of the trouble which artificially produced earth vibrations have occasioned those who make astronomical observations occurred some twenty years ago at the Greenwich Observatory. When determining the collimation error of the transit circle by means of the reflexion of a star in a tray of mercury, it was found that on certain nights the surface of the mercury was in such a state of trembling that the observers were unable to complete their observations until long after midnight. After obtaining a series of dates on which these disturbances occurred, it was found that they coincided with public and bank holidays, on which days crowds of the poorer classes of London flocked to Greenwich Park, and there amused themselves with running and rolling down the hill on which the observatory is situated. On these occasions it was found that the disturbances in the mercury were such that observations could not be made until two or three hours after the crowds had been turned out of the neighbouring park.[1]

To obviate this difficulty Sir George Airy suspended his dish of mercury in a system of indiarubber bands, and in this way succeeded in eating the intruders up.

Lieutenant-Colonel H. S. Palmer, R.E., when engaged with the transit of Venus expedition in New Zealand, in 1874, was troubled with vibrations produced from a

[1] Palmer, *Trans. Seis. Soc. of Japan*, vol. iii. p. 148.

neighbouring railway. To escape the enemy he in-
trenched his instruments by placing them in pits. With
pits 3½ feet deep he found himself sufficiently protected.
The distance from the line was about 400 yards, and the
soil through which the disturbances were propagated was
a coarse pebbly gravel.[1]

Before the United States Naval Observatory was
established at Washington, Professor H. M. Paul was
deputed to make a tremor survey to discover stable
ground. The results of these experiments were exceedingly interesting. By watching the reflected image of a
star in a dish of mercury a passing train would be noticed
at the distance of a mile. Its approach could be detected
by the trembling of the image before its coming could
be heard. At one point of observation the disturbance
appeared to be cut off by a ravine. The strata was gravel
and clay.[2]

These few examples of artificially produced tremors,
to which many more might be added, have been given
because they teach us something respecting their nature.
Hitherto earth tremors have only been regarded as
intruders, which it was necessary to escape from or des-
troy. From what has been said they appear to be a
superficial disturbance which is propagated to an enormous distance. This distance appears to depend upon
the propagating medium, upon the intensity of the
initial disturbance, and upon its duration. In the observation of these artificial disturbances, which are accessible
to every one, and which hitherto have been so neglected,
we have undoubtedly a fruitful source of study.

Natural tremors.—Next let us turn to those micro-
scopical disturbances of our soil which are due to natural

[1] Palmer, *Trans. Seis. Soc. of Japan*, vol. iii. p. 148.
[2] Paul, *Trans. Seis. Soc. of Japan*, vol. ii. p. 41.

causes. Thus far they seem to have been recorded wher-
ever instruments suitable for their detection have been
erected, and it is not improbable that they are common
to the surface of the whole globe.

Some of the more definite observations which have
been made upon earth tremors were those made in connection with experiments on the deviation of the vertical
due to the attractive influence of the moon and sun.

Professor Zöllner, who invented the horizontal pendulum which he used in the attempt to measure the change
in level due to lunar and solar attraction, found his instruments so sensitive that the readings were always changing.

The most interesting observations which were made
upon small disturbances of the soil were those of M.
d'Abbadie, who carried on his experiments at Abbadia, in
Subernoa, near Hendaye, 400 mètres distant from the
Atlantic, and 62 mètres above sea level. The soil was a
loamy rock. Here M. d'Abbadie constructed a concrete
cone 8 mètres in height, which was pierced down the
centre by a vertical hole or well, which was continued two
mètres below the cone into the solid rock. At the
bottom of this hole or well a pool of mercury was formed
which reflected the image of cross wires placed at the
top of the hole. These cross wires and their reflection
were observed by means of a microscope. The observations consisted in noting the displacement and azimuth
of the reflected image relatively to the real image of the
wires. After allowing this structure five years to settle,
M. d'Abbadie commenced his observations. To find the
surface of the mercury tranquil was a rare occurrence.
Sometimes the mercury appeared to be in violent motion,
although both the air and neighbouring sea were perfectly
calm. At times the reflected image would disappear as
if the mercury had been disturbed by a microscopic
earthquake.

The relative positions of the images were in part
governed by the state of the tide. Altogether the movements were so strange that M. d'Abbadie did not venture
any speculations as to their cause, but he remarks that
the cause of the changes he observed were sometimes
neither astronomical nor thermometrical. These observations, the principal object of which was to determine
changes in level rather than earth vibrations, were carried
on between the years 1868 and 1872.[1]

Observations at Cambridge.—Another instructive set
of observations were those which were made in the years
1880–1882, by George and Horace Darwin, in the Caven-
dish Laboratory, at Cambridge. The main object in these
experiments was to determine the disturbing influence of
gravity produced by lunar attraction. The result which
was obtained, however, showed that the soil at Cambridge
was in such an incessant state of vibration that whatever
pull the moon may have exerted upon the instrument
which was employed was masked by the magnitude of
the effects produced by the earth tremors, and the experiments had, in consequence, to be abandoned.

The principle of this instrument was similar to one
devised by Sir William Thomson, and put up by him in
his laboratory at Glasgow. As erected by the brothers
Darwin, at Cambridge, it was briefly as follows: A
pendulum, which was a massive cylinder of pure copper,
was hung by a copper wire, about four feet long, inside a
hollow cylindrical tube rising from a stone support. A
small mirror was then hung by two silk fibres, one
of which was fastened to the bob and the other to the
stone basement. A ray of light sent from a lamp on to
the mirror was reflected to a scale seven feet distant, and
by this magnification any motion of the bob relatively to

[1] G. H. and H. Darwin, *Reports of British Association*, 1881.

the stone support was magnified 50,000 times. In several ways the apparatus was insulated from all accidental disturbances. The spot of light was observed from another room by means of a telescope. This instrument was so delicate that even at the distance of sixteen feet the shifting of your weight from one foot to the other caused the spot of light to run along the scale. So sensitive was the instrument that, notwithstanding its being cut off from the surrounding soil by a trench filled with water and the whole instrument being immersed in water to damp out the small vibrations, it would seem that the ground was in a constant state of tremor ; in fact, so persistent and irregular were these movements that it seemed impossible to separate them from the perturbations due to the attraction of the moon.[1]

As a result of observations like these, the world had gradually forced upon it the fact that the ground on which we live is probably everywhere in what is practically an incessant state of vibration.

This led those who were interested in the study of earth movements to establish special apparatus for the purpose of recording these motions with the hope of eventually discovering the laws by which they were governed.

Experiments in Japan.—The simpler forms of apparatus which have been used in Japan may be described as delicate forms of seismoscopes, which, in addition to recording earth tremors, also record the occurrence of small earthquakes.

A simple contrivance which may be used for the purpose of recording small earthquakes can be made with a small compass needle.

If a light, small sensitive compass needle be placed on

[1] *Reports of British Association*, 1881.

a table, it will be found that a small piece of iron like a
nail may be pushed so near to it that the needle assumes
a position of extremely unstable equilibrium. If the
table now receives the slightest tap or shake this condition
is overcome, and the needle flies to the iron and there
remains. By making the support of the needle and the
iron the poles of an electric circuit it is possible to
register the time at which motion took place with considerable accuracy.

With crude apparatus like this a large number of
small earth disturbances have been recorded.

Another form of apparatus, employed in Japan, has
been a delicately constructed *circuit closer*. The motions
of this instrument were recorded by causing an electro-
magnet to deflect a pencil which was tracing a circle on
a revolving dial. The revolving dial was a disc of wood
covered with paper fixed to the hour-hand axle of a
common clock.

A third form of apparatus used in Japan consisted of
a small piece of sheet lead about the size of a threepenny
piece suspended by a short loop from a rigid support.
Projecting from the lead a fine wire, about two inches in
length, passed freely through a hole in a metallic plate.
By the slightest motion of the support the small pendulum
of lead was set into a state of tremor and caused its
pointer to come in contact with one or other side of the
hole in the metal plate and thus to close an electric circuit.

A more refined kind of apparatus which has been
employed in Japan was similar to that used by the
Darwins at Cambridge. This was so arranged that any
deflection of the mirror was permanent until the instrument was reset, and in this way the maximum disturbance
which had taken place between each observation was
recorded.

In addition to these and other contrivances, experiments were made with microphones.

The microphones used were small doubly pointed pencils of carbon about three centimetres long, saturated with mercury, and supported vertically in pivot holes bored in other pieces of carbon, which were the terminals of an electric circuit. These microphones were screwed down on the top of stakes driven deeply into the ground. They were covered with a glass shade thickly greased at its base. The stakes were in the ground at the bottom of a small pit—about two feet square and two feet deep—which was lined with a box. The box was covered with a lid, and earth to the depth of nine inches or one foot. One of these pits was in the middle of a lawn in the front of my house, and the other was at the foot of a hill at the back of the house. The wires from the microphone passed through the side of the box into a bamboo tube and thence up to my dining-room and bed-room. In one of the circuits there were three Daniell's cells, a telephone, and a small galvanometer. I used the galvanometer because I found that when there was sufficient motion in the microphone to produce a sound in the telephone a motion in the needle of a galvanometer was produced. If in any case motion took place in the magnetic needle during my absence, it was held deflected by a small piece of iron with which it was brought into contact by the movement.

The sensitiveness of the arrangement may be judged of from the fact that if a person walked on the grass within six feet of the microphone, each step caused a creak in the telephone, and the needle of the galvanometer was caused to swing and come in contact with the iron. Dogs running on the grass had no effect. A small stone one or two inches in diameter thrown from the house so that it fell near to the microphone pit caused a sharp creak in the telephone and a movement in the needle.

The nature of the records I received from this contrivance may be judged of from the following extract from my papers.

	h.	m.				
1879. Nov. 12th	7	0	P.M.	contact of needle		
	7	2	;,	difficult to set the needle		
	7	3	„	needle swings and telephone creaks		
	7	4	„	„.	„	„
	7	5	„	„	„	„
	7	6	„	„	„	„
	7	10	„	3 more swings		
	7	11	„	again	„	

Here I went out, took away the covering, and examined the microphone. Nothing wrong was to be observed. All that I saw was one small ant. I do not think that this could have caused the disturbance, because it could not get near the instrument.

On the succeeding nights I experienced similar disturbances, and it seemed as if they might possibly have been the prelude to several small shocks which occurred about this time (November 15, 16, and 17). On November 17, at 8 A.M., the needle was found in contact, and again at 5 P.M., and at 6 P.M. the shock of a small earthquake was felt *which caused a rattling sound in the telephone for about one minute after the motion had appeared to cease.* The needle swung considerably, but did not come in contact.

The great objection to these observations is that it is possible that the movements and sounds which I have recorded might, with the exception of one case when the shaking was actually felt, possibly have been produced by causes other than that of the movement of the ground. To determine this I subsequently put up two distinct sets of apparatus to determine whether the motions of each were synchronous. So far as I went this appeared only to be sometimes the case :—but this is a question difficult to determine, unless a recorder of time be added to the apparatus.

The greatest objection to observations of this sort is that the sensibility of the instrument is not constant. After a current has been running for several days it is no longer sensible to slight shocks, it appears as if its resistance had been increased. To overcome this it is necessary to resharpen the carbon points and bore out the pivot holes every three or four days. Farther, the battery varies. This might to some extent be overcome by using a battery with large plates. These two causes tend to reduce the sensitiveness of the galvanometer-like recorder —the deflection of the needle gradually becoming less and less, and therefore day by day needing a greater swing to bring it into contact with the iron. For reasons such as these this instrument, to be used successfully, appears to require considerable attention.

Another form of microphone employed by the author consisted of an aluminium wire standing vertically on a metallic plate, its upper end passing loosely through a hole in an aluminium wire standard.

The upper end of the vertical wire was loaded with lead. This contrivance possesses all the sensitiveness of an ordinary microphone, whilst, if it receives a sudden impulse, there is a sudden break in the current, and the vertical wire is thrown from one side to the other of the hole in the standard.

After many months of tiresome observation with instruments of this description, and after eliminating all motions which might have been produced by accidental causes, the general result obtained showed that in Tokio there were movements of the soil to be detected every day, and sometimes many times per day, which to ordinary persons were passed by unnoticed.

Work in Italy.—The most satisfactory observations which have been made upon microseismic disturbances are those which have been made during the last ten years in Italy. The father of systematical microseismical research appears to have been Father Timoteo Bertelli, of Florence.

In 1870 Father Bertelli suspended a pendulum in a cellar, and observed it with a microscope. As the result of his observations it was announced that he had perceived the earthquakes which shook Romagna, although to the ordinary observer in Florence these shakings had not been perceptible.

In 1873 Bertelli, by means of microscopes fixed in several azimuths, made 5,500 observations on free pendulums. He also made observations on reflections from the surface of mercury.[1]

One result of these observations was to show that microseismic motions increased with a fall of the barometer. Similar observations were made at Bologna by M. le Conte Malvasia, and also by M. S. di Rossi, near Rome. On January 14, 15, and February 25 these three observers at their respective stations simultaneously observed great disturbances.

Similar investigations were made at Nice by M. le Baron Prost.

Although doubt was cast upon Bertelli's observations they appear to have been the origin of a series of microseismical observations, a distinguished leader in which is Professor Rossi, who, in 1874, found that large earthquakes were almost always preceded or accompanied with micro-

---

[1] *Comptes Rendus*, 1875, January to June, p. 685.

seismical storms. In 1878 Professor Rossi worked upón these small disturbances with the assistance of the microphone and telephone, and his first results were published by Professor Palmieri.

Many of Professor Rossi's observations were made in the grotto of Roca de Papa, 700 mètres high and eighteen mètres under the soil. Here over 6,000 observations were made by means of microscopes, on pendulum's of different lengths, suspended in tubes cut in the solid rock.

Instruments employed in Italy.—It is impossible to describe in detail the various forms of apparatus which have been used by the Italian investigators. A description of one or two of the more important instruments may not, however, be out of place, inasmuch as they will assist the reader to understand the manner in which the various results respecting the laws governing microseismic movements have been arrived at.

The most important of these instruments is the *Normal Tromometer* of Bertelli and Rossi.

This consists of a pendulum $1\frac{1}{2}$ metres lóng, carrying, by means of a very fine wire, a weight of 100 grammes. To the base of the bob a vertical stile is attached, and the whole is enclosed in a tube terminated, at its base, by a glass prism of such a form that when looked through horizontally the motion of the stile can be seen in all azimuths. In front of this prism a microscope is placed. Inside the microscope there is a micromatic scale, so arranged that it can be turned to coincide with the apparent direction of oscillation of the point of the stile. In this way not only can the amplitude of the motion of the stile be measured, but also its azimuth. The extent of vertical motion is measured by the up and down motion of the stile due to the elasticity of the supporting wire. This instrument is shown in the accompanying drawing.

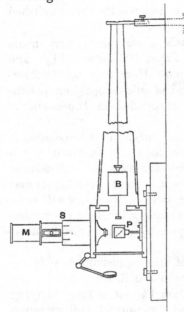

FIG. 37.—Normal Tromometer.
B, bob of pendulum; P, prism; M, microscope;
S, rim of scale.

Another apparatus is the *Microseismograph* of Professor Rossi. Here we have an arrangement which gives automatic records of slight motions. It consists of four pendulums, each about three feet long, suspended so that they form the corners of a square platform. In the centre of this platform a fifth, but rather longer, pendulum is suspended. The four pendulums are each connected just above their bobs to the central pendulum with loose silk threads. Fixed to the centre of each of these threads, and held vertically by a light spring, is a needle, so adjusted that each thread is depressed to form an obtuse angle of about 155°. These needles form the terminals of an electric circuit, the other termination of which is a small cup of mercury placed just below the lower end of the needle. By a horizontal swing of one of the pendulums this arrangement causes the needle to move vertically, but with a slightly multiplied amplitude. By this motion the needle comes in contact with the mercury, and an electro magnet with a lever and pencil is caused to make a mark on a band of paper moved by clockwork. The five pendulums being of different lengths allows the apparatus to respond 'to seismic waves of different velocities.'[1]

Lastly, we have Professor Rossi's *Microphone*. This consists of a metallic swing arranged like the beam of a balance. By means of a moveable weight at one end of the beam this is so adjusted that it falls down until it comes in contact with a metallic stop. This can be so adjusted that a slight tap will cause the beam to slightly jump from the stop. The beam and the stop form two poles of an electric circuit, in which there is a telephone. The slightest motion in a *vertical* direction causes a fluctuation in the current passing between the stop and the beam, and a consequent noise is heard in the telephone.

With instruments analogous to these, observations have been made by various observers in all portions of Italy, extending over a period of ten years. Every precaution appears to have been taken to avoid accidental disturbances, and the experiments have been repeated in a variety of forms.

Results obtained in Italy.—The results which from time to time have been announced are of the greatest interest to those who study the physics of the earth's crust, and they appear to be leading to the establishment of laws of scientific value.

It would seem that the soil of Italy is in incessant movement, there being periods of excessive activity usually lasting about ten days. Such periods are called seismic storms. These storms are separated by periods of relative calm. These storms have their greater regularity in winter, and sharp maximums are to be observed in spring and autumn. In the midst of such a period or

[1] *Tel. Journ.*, November, 15, 1881.

at its end there is usually an earthquake. Usually these
storms are closely related to barometric depressions. To
distinguish these movements from those which occur
under high pressure, the latter are called *baro seismic*
movements, and the former *vulcano seismic* movements.
The relation of these storms to barometric fluctuation has
been observed to have been very marked during the time
of a volcanic eruption.

At the commencement of a storm the motions are
usually small, and one storm, lasting two or three days,
may be joined to another storm. In such a case the
action may be a local one. It has been observed that a
barometrical depression tended to bring a storm to a
maximum, whilst an increase of pressure would cause it
to disappear. Sometimes these actions are purely local,
but at other times they may affect a considerable tract of
land.

If a number of pendulums of different length are
observed at the same place, there is a general similarity
in their movements, but it is also evident that the free
period of the pendulum more or less disturbs the character
of the record. The greatest amplitude of motion in a set
of pendulums is not reached simultaneously by all the
pendulums, and at every disturbance the movement of
one will predominate. From this Rossi argues that the
character of the microseismical motions is not constant.
Bertelli observed that the direction of oscillation of the
pendulums is different at different places, but each place
will have its particular direction dependent upon the
direction of valleys and chains of mountains in the neighbourhood. Rossi shows that the directions of movement
are perpendicular to the direction of lines of faults, the
lips of these fractures rising and falling, and producing
two sets of waves, one set parallel to the line of fracture,

and the other perpendicular to such a direction. These movements, according to Bertelli, have no connection with the wind, rain, change of temperature, and atmospheric electricity.

The disturbances, as recorded at different towns, are not always strictly synchronous, but succeed each other at short intervals. If, however, we take monthly curves of the disturbances as recorded at different towns in Italy,

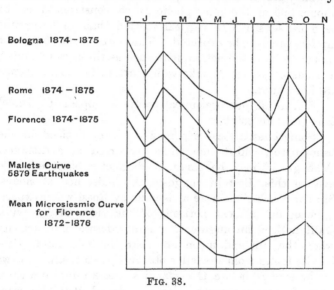

Bologna 1874 – 1875

Rome 1874 – 1875

Florence 1874 - 1875

Mallets Curve
5879 Earthquakes

Mean Microsiesmic Curve
for Florence
1872 – 1876

FIG. 38.

we see that these are similar in character. The maximum of disturbance occurs about the winter solstice, and the minimum about the summer solstice, and in this respect they exhibit a perfect accordance with the curves drawn by Mallet to show the periodicity of earthquakes. The accompanying curves taken from one of Bertelli's original memoirs not only show this general result but also show

Y

the close accord there is between the results obtained at different towns during successive months.

At Florence, before a period of earthquakes there is an increase in the amplitude and frequency of vertical movements. These vertical movements do not appear to coincide with the barometrical disturbances, but they appear to be connected with the seismic disturbances.

They are usually accompanied with noises in the telephone, but as the microphone is so constructed as to be more sensitive to vertical motion than to horizontal motion, this is to be expected. This vertical motion would appear to be a local action, inasmuch as the accompanying motions of an earthquake which originates at a distance are horizontal.

Storms of microseismical motions appear to travel from point to point.

Sometimes a local earthquake is not noticed in the tromometer, whilst one which occurred at a distance, although it may be small, is distinctly observed. To explain this, Bertelli suggests the existence of nodes. Similar conclusions were arrived at by Rossi when experimenting on different portions of the sides of Vesuvius. Galli noticed an augmentation in microseismic activity when the sun and moon are near the meridian. Grablovitz found from Bertelli's observations a maximum two or three days before the syzygies, and a minimum three days after these periods. He also found that the principal large disturbances occurred in the middle of periods separating the quadrature from the syzygies, the apogee from perigee, and the lunistigi period from the nodes, whilst the smallest disturbances happened in the middle of periods opposed to these.

P. C. Melzi says that the curves of microseismical motions, earthquakes, lunar and solar motions, show a concordance with each other.

With the microphone Rossi hears sounds which he describes as roarings, explosions, occurring isolated or in volleys, metallic and bell-like sounds, ticking, &c., which, he says, revealed natural telluric phenomena. Sometimes these have been intolerably loud. At Vesuvius the vertical shocks corresponded with a sound like volleys of musketry, whilst the undulatory shocks gave the roaring. Some of these sounds could be imitated artificially by rubbing together the conducting wires in the same manner in which the rocks must rub against each other in an earthquake. Other sounds were imitated by placing the microphone on a vessel of boiling water, or by putting it on a marble slab and scratching and tapping the under side of it.

These, then, are some of the more important results which have been arrived at by the study of microseismic motions. One point which seems worthy of attention is that they appear to be more law-abiding than their more, violent relations, the earthquakes, and as phenomena in which natural laws are to be traced they are certainly deserving our attention. As to whether they will ever become the means of forewarning ourselves against earthquakes is yet problematical. Their systematic study, however, will enable us to trace the progress of a microseismic storm from point to point, and it is not impossible that we may yet be enabled to foretell where the storm may reach its climax as an earthquake. These, I believe, are the views of Professor di Rossi, who is at the present time engaged in the establishment of a system of microseismic observations throughout Italy.

Before the earthquake of San Remo (Dec. 6, 1874) Rossi's tromometer was in a state of agitation, and similar disturbances were observed at Livorno, Florence, and Bologna.

Y 2

Since February 1883 I have observed a tromometer in Japan, and such results as have been obtained accord with results obtained in Italy. The increase in micro-seismical activity with a fall of the barometer is very marked. Other peculiarities in the behaviour of the instrument will be referred to under 'Earth Pulsations.'

Cause of microseismic movements.—As to the cause of tromometric movements, we have a field for speculation. Possibly they may be due to slight vibratory motions produced in the soil by the bending and crackling of rocks produced by their rise upon the relief of atmospheric pressure. If this were so we should expect similar movements to be produced at the time of an increase of pressure. Rossi suggests that they may be the result of an increased escape of vapour from the molten materials beneath the crust of the earth, consequent upon a relief of pressure. The similarity of some of the sounds which are heard with the microphone to those produced by boiling water are suggestive of this, and Rossi quotes instances when underground noises like those which we should expect to hear from a boiling fluid have been heard before earthquakes without the aid of microphones. One instance was that of Viduari, a prisoner in Lima, who, two days before the shock of 1824, repeatedly predicted the same in consequence of the noises he heard.

A possible cause of disturbances of this order may be small but sudden fluctuations in barometric pressure, which are visible during a storm. During a small typhoon on September 15, 1881, when in the Kurile Islands, I observed that the needle of an aneroid worked back and forth with a period of from one to three seconds. This continued for several hours. With every gust of wind the needle suddenly rose and then immediately fell. At times it trembled. These movements were observed in the open air. The extent of these sudden variations was approximately from ·03 to ·05 inches. Reckoning an increase of barometrical pressure of one hundredth of an inch as equivalent to a load of twenty million pounds on the square mile, during this storm there must have been the equivalent of loads of from 60 to 100 million pounds to the square mile continually placed on and removed from a considerable tract of the earth's surface. If the period of application of these stresses approximately coincide with the natural vibrational period of the area affected, it would surely seem, especially when we reflect upon the effect of an ordinary carriage, that tremors of considerable magnitude ought to be produced.

An inspection of the following few observations taken from my note-book for the same typhoon will suggest that even the large and slower variations are capable of producing tremulous motions.

Time h. m.								Barometer reading
12 5	P.M.	.	.	.	.	.	.	29·02
12 10	,,	.	.	.	.	.	.	29·05
12 12	,,	.	.	.	.	.	.	29·07
12 13	,,	.	.	.	.	.	.	29·05
12 25	,,	.	.	.	.	.	.	29·10
12 50	,,	.	.	.	.	.	.	29·00
1 10	,,	.	.	.	.	.	.	29·00
1 20	,,	.	.	.	.	.	.	29·07

CHAPTER XX.

EARTH PULSATIONS.

Definition of an earth pulsation—Indications of pendulums—Indications of levels—Other phenomena indicating the existence of earth pulsations—Disturbances in lakes and oceans—Phenomena resultant on earth pulsations—Cause of earth pulsations.

THE object of the present chapter is to show that from time to time it is very probable that slow but large wave-like undulations travel over or disturb the surface of the globe.

These movements, which have escaped our attention on account of their slowness in period, for want of another term I call earth·pulsations.

The existence of movements such as these may be indicated to us by changes in the level of bodies of water like seas and lakes, by the movements of delicate levels, by the displacement of the bob of a pendulum relatively to some point on the earth above which it hangs, and by other phenomena which will be enumerated.

Indication of pendulums.—Pendulums which have been suspended for the purposes of seismometrical obser-vations have, both by observers in Italy and Japan, been seen to have moved a short distance out from, and then back to, their normal position.

This motion has simply taken place on one side of their central position, and is not due to a swing. The

character of these records is such that we might imagine
the soil on which the support of the pendulum had rested
to have been slowly tilted, and slowly lowered. They are
the most marked on those pendulums provided with an
index writing a record of its motions on a smoked glass
plate, which index is so arranged that it gives a multiplied representation of the relative motion between it and
the earth. As motions of this sort might be possibly due
to the action of moisture in the soil tilting the support
of the pendulum, and to a variety of other accidental
causes, we cannot insist on them as being certain indications that there are slow tips in the soil, but for the
present allow them to remain as possible proof of such
phenomena.

Evidence of displacement of the vertical, which are
more definite than the above, are those made by Bertelli,
Rossi, Count Malvasi, and other Italian observers, who,
whilst recording earth tremors, have spent so much time
in watching the vibrations of stiles of delicate pendulums
by means of microscopes. As a result of these observations we are told that the point about which the stile of
a pendulum oscillates is variable. These displacements
take place in various azimuths, and they appear to be
connected with changes of the barometer. I have made
similar observations in Japan.

From this, and from the fact that it is found that a
number of different pendulums differently situated on
the same area give similar evidence of these movements,
it would hardly seem that these phenomena could be
attributed to causes like changes in temperature and
moisture. M. S. di Rossi lays stress on this point, especially
in connection with his microseismograph, where there are
a number of pendulums of unequal length which give
indications of a like character. The direction in which

these tips of the soil take place—which phenomena are noticeable in seismic as well as microseismic motions—Rossi states are related to the direction of certain lines of faulting.

Indications of levels.—Bubbles of delicate levels can be easily seen to change their position with meteorological variations; but Rossi also tells us that they change their position, sometimes not to return for a long time, during a microseismic storm. Here again we have another phenomenon pointing to the fact that microseismic disturbances are the companions of slow alterations in level.

One of the most patient observers of levels has been M. Plantamour, who commenced his observations in 1878, at Sécheron, on the Lake of Geneva. He used two levels, one placed north and south, and the other east and west. During the summer of 1878 the east end rose, but at the end of September a depression set in. The diurnal movements had their maximum and minimum at 6 and 7.45 A.M. and P.M. The total amplitude was 4·89″. The variations of the east and west level appeared to be due to the temperature, but the movements of the north and south level were dependent upon an unknown cause.

Between October 1, 1879, and September 30, 1880, the east end fell rapidly, from the middle of November up to December 26, amounting to 88·71″. It then rose 6·55″ to January 5, and then fell again. On January 28 it reached 89·95″, after which it rose.

Between October 4, 1879, and January 28, 1880, the movement was 95·8″, against 28·08″ of the previous year.

These movements were not due alone to temperature. The north and south level, which was not influenced by the cold of the winter, moved 4·56″. In the previous year 4·89″.[1]

From February 17 to June 5, 1883, the author observed in Tokio the bubbles of two delicate levels, one placed north and south, and the other east and west. They were placed under glass cases on the head of a stone column. The column, which is inside a brick building, rests on a concrete foundation, and is about ten years old. It is in no way connected with the building. The temperature of the room has a daily variation of about 1° Fahr.

In both these levels diurnal changes are very marked. Occasionally they are enormously great. Thus, on March 25, the readings of the south end of the north south level were as follows :—

Time h. m.						Readings.
25th. 4 00 P.M.	.	.	.	.		104·5
4 5 ,,	.	.	.	.		103
4 10 ,,	.	.	.	.		102
4 25 ,,	.	.	.	.		101
4 30 ,,	.	.	.	.		100
4 40 ,,	.	.	.	.		98
4 42 ,,	.	.	.	.		99·5
4 45 ,,	.	.	.	.		100
4 50 ,,	.	.	.	.		101
4 55 ,,	.	.	.	.		101
5 00 ,,	.	.	.	.		100
26th. 7 00 A.M.	.	.	.	.		105

Usually this level moves through about three divisions per day.

From March 25 to May 4 it travelled from 98 to 127. Since then, to June 5, it has descended to 116. During this period the east west level has been *comparatively* quiet. One division of the north south level equals about 2″ of arc.

---

[1] *Minutes and proceedings of the Institute of Civil Engineers*, vol. lx. p. 412, and vol. lxiv. p. 343.

Many of these changes may be due to changes in temperature, variations in moisture, and other local actions. Some of them, however, are hardly explicable on such assumptions. The fact that the general direction in change of the vertical, as indicated by a tromometer standing on the same column with the levels, showed that the change which was taking place was rather in the column than in the instruments.

The fact also that at the time of a barometrical depression a *pulse-like surge* can be seen in the levels, having a period averaging about three seconds and sometimes amounting to about one second of arc, is a phenomenon hardly to be attributed to sudden fluctuations in moisture or temperature, but indicates real changes in level.[1]

In addition to variation in the bubbles of levels which come on more or less gradually, we have many recorded instances of *sudden* alterations taking place in these instruments.

Examples of what may have been a slow oscillating motion of the earth's crust are referred to by Mr. George Darwin in a Report to the British Association in 1882.

One of them was made by M. Magnus Nyrén, at Pulkova, who, when engaged in levelling the axis of a telescope, observed spontaneous oscillation in the bulb of the level.

This was on May 10 (April 28), 1877. The complete period was about 20 seconds, the amplitude being 1·5″ and 2″. One hour and fourteen minutes before this he observes that there had been a severe earthquake at Iquique, the distance to which in a straight line was 10,600 kilomètres, and on an arc of a great circle, 12,500 kilomètres. On September 20 (8), in 1867, Mr. Wagner

[1] *See* 'Earth Tremors,' p. 309, experiments of M. d'Abbadie, &c.

had observed at Pulkova oscillations of 3″, seven minutes before which there had been an earthquake at Malta. On April 4 (March 23), 1868, an agitation of the level had been observed by Mr. Gromadzki, five minutes before which there had been an earthquake in Turkestan. Similar observations had been made twice before. These, however, had not been connected with any earthquakes— at least, Mr. Darwin remarks—with certainty.

Phenomena analogous to the pendulum and level observations.—As examples of phenomena which are analogous to those made on pendulums and levels, the following may be noticed. On March 20, 1881, at 9 P.M. a watchmaker in Buenos Ayres observed that all his clocks oscillating north and south suddenly began to increase their amplitude, until some of them became twice as great as before. Similar observations were made in all the other shops. No motion of the earth was detected. Subsequently it was learnt that this corresponded with an earthquake in Santiago and Mendoza.[1]

Another remarkable example illustrating the like phenomena is furnished by the observations which were made on December 21, 1860, by means of a barometer in San Francisco, which oscillated, with periods of rest, for half an hour. No shock was felt, nor is it likely that it was a local accident, as it could not be produced artificially. On the following day, however, a violent earthquake was experienced at Santiago.[2]

At the time or shortly after the great Lisbon earthquake, curious phenomena were observed in distant countries, which only appear to be explicable on the assumption of the existence of earth pulsations.

Thus at Amsterdam and other towns, chandeliers in churches were observed to swing. At Haarlem water was

[1] *Meteorologia Endogena.*   [2] *Ibid.*

thrown over the sides of tubs, and it is expressly mentioned
that no motion was perceived in the ground.

At the Hague a tallow chandler was surprised at the
clashing noise made by his candles, and this the more so
because no motion was felt underfoot.

Unusual disturbances in bodies of water.—At the
time of large earthquakes it would appear that earth
pulsations are produced, which exhibit themselves in
countries where the actual shaking of the earthquake is
not felt, by disturbances in bodies of waters like lakes and
seas.

Some remarkable examples of these disturbances are
to be found in the records of the great Lisbon earthquake.
This earthquake, as a violent movement of the ground,
was chiefly felt in Spain, Portugal, northern Italy, the
south of France and Germany, northern Africa, Madeira,
and other Atlantic islands. In other countries further
distant, as, for instance, Great Britain, Holland, Scandinavia, and North America, although the records are
numerous, the only phenomena which were particularly
observed was the slow oscillations of the waters in lakes,
ponds, canals, &c. In some instances the observers
especially remark that there was no motion in the
soil.

Pebley Dam, in Derbyshire, which is a large body of
water covering some thirty acres, commenced to oscillate
from the south. A canal near Godalming flowed eight
feet over the walk on the north side.

Coniston Water, in Cumberland, which is about five
miles long, oscillated for about five minutes, rising a yard
up its shores. Near Durham a pond, forty yards long
and ten broad, rose and fell about one foot for six or seven
minutes. There were four or five ebbs and flows per
minute.

Loch Lomond rose and fell through about two and a half feet every five minutes, and all the other lochs in Scotland seem to have been similarly agitated.

At Shirbrun Castle, in Oxfordshire, where the water in some moats and ponds was very carefully observed, it was noticed that the floods began gently, the velocity then increased, till at last with great impetuosity they reached their full height. Here the water remained for a little while, until the ebb commenced, at first gently, but finally with great rapidity. At two extremities of a moat about 100 yards long, it was found that the sinkings and risings were almost simultaneous. The motions in a pond a short distance from the moat were also observed, and it was found that the risings and sinkings of the two did not agree.

During these motions there were several maxima.

These few examples of the motions of waters, without any record of the motions of the ground, at the time of the Lisbon earthquake, must be taken as examples of a very large number of similar observations of which we have detailed accounts.

Like agitations, it must also be remembered, were perceived in North America and in Scandinavia, and if the lakes of other distant countries had been provided with sufficiently delicate apparatus, it is not unlikely that similar disturbances would have been recorded.

Besides these movements in the waters of seas and lakes, at or about the time of great earthquakes, we have records of like movements, which take place as independent phenomena.

Thus we read that on October 22, 1755, the waters of Lake Ontario rose and fell five and a half feet several times in the course of half an hour.[1] On March 31, 1761,

[1] *Phil. Trans.* vol. xlix. p. 544.

Loch Ness rose suddenly for the period of three quarters of an hour.[1]

As another example of the disturbance of water at the time of a great earthquake in districts where the earthquake was not felt, may be mentioned the swelling of the waters of the Marañon, in 1746, on the night when Callao was overwhelmed.

Sudden variations in the level of the water have been many times observed in the North American lakes. The changes in level which sometimes take place in the Genfer and Boden lakes are supposed to have some relation to the condition of the atmosphere. A rising and falling of especial note took place on April 18, 1855.

In Switzerland these sudden changes are known as 'seiches' or 'rhussen.'

From the observations and calculations of Prof. Forrel it would seem that the period of the 'seiches' depends upon the dimensions of the lakes ; the calculated periods dependent on the depths of the lakes being approximately equal to the observed periods.[2]

W. T. Bingham, writing on the volcanoes of the Hawaian Islands, remarks that it is not unusual for the sea to be agitated by great and unusual tides, and that such sea waves have not been attended with volcanic eruptions or seismic disturbances. Thus in May 1819 the tide rose and fell thirteen times. On November 7, 1837, there was an ebb and flow of eight feet every twenty-eight minutes. Again, on May 17, 1841, like phenomena, unaccompanied by any other unusual occurrences, were recorded.[3]

Phenomena which may possibly hold a relationship to earth pulsations are the periodical swellings of the ocean on the coast of Peru. Dr. C. F. Winslow, who made a long

[1] *Annual Register*, vol. iv. 1761, p. 92.
[2] *Phil. Mag.*, May 1876, p. 447.   [3] *Boston Soc. Nat. Hist.*, 1868.

period observation upon the coast of Peru, found 'the highest tides to prevail at Callao and Paita in December and January,' and 'also a series of enormous waves or sea-swells to be thrown from time to time upon the coast, varying from twenty-four to twenty-seven hours in continuance, accompanied by unusual height of the tide during the same period.' During June and July the ocean was unusually tranquil. These phenomena do not appear to be connected with great atmospheric storms, nor do they hold any relation to the prevailing wind. They increase with and accompany the swelling of the tides, and occur generally, but not always, about full moon.

Sometimes they break suddenly upon the coast. *They are annual and constant in their periodicity.*'

The periodical swellings are most noticeable between Tumbez 3° S.L. and the Chincha Islands 14° S.L.

These oceanic phenomena synchronise with the periodic intensity of earthquake phenomena in that part of the globe, and these with tidal movements.[1]

Other phenomena possibly attributable to earth pulsations.—If we assume that earth- pulsations have an existence, these many phenomena which are otherwise difficult to understand meet with an explanation. The curious effects which were produced in the springs at Toplitz at the time of the Lisbon earthquake may have been due to a pulse-like wave. The flow of the principal spring was greatly increased. Before the increase it became turbid and at one time stopped. Subsequently it became clear and flowed as usual, but the water was hotter and more strongly mineralised. Sudden changes in the flow of underground waters which from time to time are observed may be attributed to like causes. Secondary earthquakes such as occurred after the Lisbon

[1] 'Notes on Tides at Tahiti' &c., *Am. Jour Sci.* 1866, vol. xlii. p. 45.

earthquake, as for instance in Derbyshire, may have been produced by pulsations disturbing the equilibrium of ground in a critical state.

The falling in of subterranean excavations is also possibly connected with these phenomena.

Possible causes of earth pulsations.—Mr. George Darwin, in a report to the British Association (1882), has shown that movements of considerable magnitude may occur in the earth's crust in consequence of fluctuations in barometrical pressure. (A rise of the barometer over an area is equivalent to loading that area with a weight, in consequence of which it is depressed. When the barometer falls, the load is removed from the area, which, in virtue of its elasticity, rises to its original position. This 'fall and rise of the ground completes a single pulsation.)

On the assumption that the earth has a rigidity like steel, Mr. Darwin calculates that if the barometer rises an inch over an area like Australia, the load is sufficient to sink that continent two or three inches.

The tides which twice a day load our shores cause the land to rise and fall in a similar manner. On the shores of the Atlantic, Mr. Darwin has calculated that this rise and fall of the land may be as much as 5 inches. By these risings and fallings of the land the inclination of the surface is so altered that the stile of a plummet suspended from a rigid support ought not always to hang over the same spot. There would be a deflection of the vertical.

In short, calculations respecting the effects of loads of various descriptions, which we know are by natural operations continually being placed upon and removed from the surface of various areas of the earth's surface, indicate that slow pulsatory movements of the earth's surface must be taking place, causing variations in inclination of one portion of the earth's crust relatively to another.

Although it is possible that phenomena like the surging of levels may be attributable to causes like these, we can hardly attribute the other phenomena to such agencies.

Rather than seek an explanation from agencies exogenous to our earth, we might perhaps with advantage appeal to the endogenous phenomena of our planet. When the barometer falls, which we have shown corresponds to an upward motion of the earth's crust, we know, from the results of experiments, that microseismic motions are particularly noticeable.

As a pictorial illustration of what this really means, we may imagine ourselves to be residing on the loosely fitting lid of a large cauldron, the relief of the external pressure over which increases the activity of its internal ebullition—the jars attendant on which are gradually propagated from their endogenous source to the exterior of our planet. This travelling outwards would take place much in the same way that the vibrations consequent to the rattle and jar of a large factory slowly spread themselves farther and farther from the point where they were produced.

Admitting an action of this description to take place, it would then follow that this extra liberation of gaseous material beneath the earth's crust would result in an increased upward pressure from within, and a tendency on the part of the earth's crust to elevation. If we accept this as an explanation of the increased activity of a tremor indicator, then such an instrument may be regarded as a barometer, measuring by its motions the variations in the internal pressure of our planet.

z

The relief of external pressure and the increase of internal pressure, it will be observed, both tend in the same direction—namely, to an elevation of the earth's crust.

This explanation of the increased activity of earth tremors, which has also been suggested by M. S. di Rossi, is here only advanced as a speculation, more probable perhaps than many others.

We know how a mass of sulphur which has been fused in the presence of water in a closed boiler gives up in the form of steam the occluded moisture upon the relief of pressure. In a similar manner we see steam escaping from volcanic vents and cooling streams of lava. We also know how gas escapes from the pores and cavities in a seam of coal on the fall of the barometrical column. We also know that certain wells increase the height of their column under like conditions. The latter of these phenomena, resulting in an increase in the rate of drainage of an area by its tendency to render such an area of less weight, facilitates its rise. If we follow the views of Mr. Mallet in considering that the pressures exerted on the crust of our earth may in volcanic regions be roughly estimated by the height of a column of lava in the volcanoes of such districts, we see that in the neighbourhood of a volcano like Cotopaxi the upward pressures must be enormously great. Further, the phenomena of earthquakes and volcanoes indicate that these pressures are variable. Before a volcano bursts forth we should expect that there would be in its vicinity an upward bulging of the crust, and after its formation a fall. Further, it is not difficult to conjecture other possible means by which such pressures may obtain relief.

Should these pressures then find relief without rupturing the surface, it is not difficult to imagine them as the originators of vast pulsations which may be recorded on the surface of the earth as wave-like motions of slow period.

As an explanation of the strange movements observed on seas and lakes, Kluge brings forward the following strange and remarkable theory. The oxygen of the air is magnetic, whilst water is diamagnetic and the earth magnetic: we have, therefore, in our seas and lakes a diamagnetic body lying between and being, consequently, repelled from two magnetic bodies. By variation in temperature, the balance of repulsions exerted by the air and the earth is destroyed. Thus, by an elevation of temperature the air expands and flows away from the heated area, where, in consequence, there is less oxygen. The result of this is, that the repulsion of the air upon the waters is less than that of the earth upon the waters, and the waters are in consequence raised up. By a falling of temperature the waters may be depressed, and by either of these actions waves may be produced without the intervention of earthquakes or earth pulsations.

The more definite kinds of information which we have to bring forward, tending to prove the existence of earth pulsations, too slow in period to be experienced by ordinary observers, are those which appear to be resultant phenomena of great earthquakes.

The phenomena that we are certain of in connection with earth vibrations, whether these vibrations are produced artificially by explosions of dynamite in bore-holes, or whether they are produced naturally by earthquakes, are, first, that a disturbance as it dies out at a given point often shows in the diagrams obtained by seismographs a decrease in period; and, secondly, a similar decrease in the period of the disturbance takes place as the disturbance spreads.

As examples of these actions I will quote the following.

The diagram of the disturbance of March 1, 1882, taken at Yokohama, shows that the vibrations at the commencement of the disturbance had a period of about three per second, near the middle of the disturbance the period is about 1·1, whilst near the end the period has decreased to ·46. That is to say, the backward and forward motion of the ground at the commencement of the earthquake was six times as great as it was near the end, when to make one complete oscillation it took between two and three seconds. Probably the period became still less, but was not recorded owing to the insensibility of the instruments to such slow motions.[1]

We have not yet the means of comparing together diagrams of two or more earthquakes, one having been taken near to the origin, and the other at a distance. The only comparisons which I have been enabled to make have been those of diagrams taken of the same earthquake, one in Tokio and the other in Yokohama. As this base is only sixteen miles, and the earthquake may have originated at a distance of several hundreds of miles, comparisons like these can be of but little value.

Other diagrams illustrating the same point are those obtained at three stations in a straight line, but at different distances from the origin of a disturbance produced by exploding a charge of dynamite in a bore-hole. A simple inspection of these diagrams shows that at the near station the disturbance consisted of backward and forward motions, which, as compared with the same disturbance as recorded at a more distant station, were very rapid. Further, by examining the diagram of the motions,

[1] *Trans. Seis. Soc. of Japan*, vol. iv. Milne, *Systematic Observation of Earthquakes.*

say, at the near station, it is clearly evident that the period of the backward and forward motion rapidly decreased as the motion died out.

These illustrations are given as examples out of a large series of other records, all showing like results.

An observation which confirms the records obtained from seismographs respecting the increase in period of an earthquake as it dies out I have had opportunities of twice making with my levels. After all perceptible motion of the ground subsequent upon a moderately severe shock had died away, I have distinctly seen the bubble in one of these levels slowly pulsating with an irregular period of from one to five seconds.

Although we must draw a distinction between earth waves and water waves, we yet see that in these points they present a striking likeness. Let us take, for example, any of the large earthquake waves which have originated off the coast of South America, and then radiated outwards, until they spread across the Pacific, to be recorded in Japan and other countries perhaps twenty-five hours afterwards, at a distance of nearly 9,000 miles from their origin. Near this origin they appeared as walls of water which were seen rapidly advancing towards the coast. These have been from twenty to two hundred feet in height, and they succeeded each other at rapid intervals, until finally they died out as a series of gentle waves. By the time these walls of water traversed the Pacific, to, let us say, Japan, they broadened out to a swell so flat that it could not be detected on the smoothest water excepting along shore lines where the water rose and fell like the tide. Instead of a wall of water sixty feet in height, we had long flat undulations perhaps eight feet in height, but with a distance from crest to crest of from one to two hundred miles.

If we turn to the effects of large earthquakes as
exhibited on the land, I think that we shall find records
of phenomena which are only to be explained on the
assumption of an action having taken place analogous
to that which takes place so often in the ocean, or an
action similar to that exhibited by small earthquakes,
and artificially produced disturbances, if greatly exaggerated.

The only explanation for the phenomena accompanying
the Lisbon earthquake appears to be that the short quick
vibrations which had ruined so many cities in Portugal
had, by the time that they had radiated to distant countries,
gradually become changed into long flat waves having a
period of perhaps several minutes. In countries like
England these pulse-like movements were too gentle to be
perceived, except in the effects produced by tipping up
the beds of lakes and ponds.

The phenomenon was not unlike that of a swell produced by a distant storm. It would seem possible that
in some cases pulsations producing phenomena like the
'seiches' of Switzerland might have their origin beneath the
ocean, or deep down beneath the earth's crust.   Perhaps,
instead of commencing with the 'snap and jar' of an
earthquake, they may commence as a heaving or sinking
of a considerable area, which may be regarded as an uncompleted effort in the establishment of an earthquake or
a volcano.

From what has now been said it would seem that
earth pulsations are phenomena with a real existence, and
that some of these are attributable to earthquakes. On
the other hand, certain earthquakes are attributable to
earth pulsations. Some of the phenomena which have
been brought forward have only a possible connection
with these movements, and they yet require investigation.

CHAPTER XXI.

EARTH OSCILLATIONS.

Evidences of oscillation—Examples of oscillation—Temple of Jupiter
Serapis—Observations of Darwin—Causes of oscillation.

Evidences of oscillation.—By earth oscillations are
meant those slow and quiet changes in the relative level
of the sea and land which geologists speak of as elevations or subsidences. These movements are especially
characteristic of volcanic and earthquake-shaken countries.

As evidences of elevations we appeal to phenomena
like raised beaches, sea-worn caves, raised coral reefs, and
the remains of other dead organisms like barnacles, and
the borings of lithodomous shells in and on the rocks of
many coasts high above the level of the highest tides.
As a proof that subsidence has taken place, there is the
evidence afforded by submerged forests, the prolongation
of certain valleys beneath the bed of the ocean, the
formation of coral islands, the peculiar distribution of the
plants and animals which we find in many countries, and
the submergence of works of human construction. Inasmuch as these phenomena are discussed so fully in many
treatises on physical geology, the references to them here
will be made as brief as possible. Elevations and depressions which have taken place at the time of large earth-

quakes in a paroxysmal manner have already been mentioned. The movements referred to in this chapter, although generally taking place with extreme slowness, in
certain instances, by an increase in their rapidity, have
approached in character to earth pulsations. In most
instances it would appear that the upward movement of
the ground, which may be likened to a process of tumefac-
tion, goes on so gently that it only becomes appreciable
after the lapse of many generations.

Examples of movements.—Lyell estimated that the
average rate of rise in Scandinavia has been about two and a
half feet per century. At the North Cape the rise may have
been as much as five or six feet per century. Observations
made at the temple of Jupiter Serapis, between October
1822 and July 1838, showed that the ground was sinking
at the rate of about one inch in four years. Since the
Roman period, when this temple was built, the ground has
sunk twenty feet below the waves. Now the floor of the
temple is on the level of the sea. Lyell remarks that if
we reflect on the dates of the principal oscillations at this
place there appears to be connection between the movements of upheaval and a local development of volcanic
heat, whilst periods of depression are concurrent with
periods of volcanic quiescence.[1]

As examples of movements even more rapid than those
at the Temple of Jupiter Serapis we refer to an account
of the earthquakes in Vallais (November 1755), when
the ground about a mountain at a small distance from
Brigue sank about a thumb's-breadth every twenty-four
hours. This took place between December 9 and February 26.[2]

Another remarkable example of earth movement is

[1] *Principles of Geology*, vol. ii. 177.
[2] *Gent. Mag.*, vol. xxvii. p. 448.

given in the account of the earthquake at Scarborough, on December 29, 1737, when the head of the spa water well was forced up in the air about ten yards high. At this time the sands on the shore are said to have risen so slowly that people came out to watch them.[1]

Two other examples of rapid earth movement are taken from Professor Rossi's 'Meteorologia Endogena.' Professor D. Seghetti, writing to Professor Rossi, says that a few lustres ago (one lustre = twenty years) Mount S. Giovanni hid the towns Jenne and Subiaco from each other. From Subiaco the church at Jenne is now visible, which a few years ago was invisible. The people at Jenne also can see more than formerly. The supposition is that the side of Mount S. Giovanni is lowered. This fact corresponds to a fact stated by Professor Carina, who says that forty or fifty years ago from Granaiola you could not see either the church of S. Maria Assunta di Citrone or the church of S. Pietro di Corsena. Now you can see both.[2]

For a remarkable example illustrating the connection between seismic activity and elevation we are indebted to the patient labours of Darwin, who carefully investigated the evidences of elevation which are visible upon the western coasts of South America. These evidences, consisting of marks of erosion, caves, ancient beaches, sand dunes, terraces of gravel, &c., were traced between latitudes 45° 35' to 12° 5', a distance north and south of 2,075 geographical miles, and there is but little doubt that they extend much farther. As deduced from observations upon upraised shells alone, a summary of Mr. Darwin's observations are contained in the following table :—

[1] *Phil. Trans.*, vol. xli. p. 805.
[2] *Meteorologia Endogena*, vol. i. pp. 186, 187.

						Feet
At Chiloe the recent elevation has been			.	.	.	350
„ Concepcion	„	„	.	.	625 to	1,000
„ Valparaiso	„	„	.	.	.	1,300
„ Coquimbo	„	„	.	.	.	252
„ Lima	„	„	.	.	.	85

Shells, similar to those clinging to uplifted rocks, which are evidences of these elevations, still exist in the neighbouring seas, and in the same proportionate numbers as they are found in the upraised beds. In addition to this, Mr. Darwin shows us that at Lima, during the Indo-human period, the elevation has been at least eighty-five feet. At Valparaiso, during the last 220 years, the rise was about nineteen feet, and in the seventeen years subsequent to 1817 the rise has been ten or eleven feet, a portion only of which can be attributed to earthquakes. In 1834 the rise there was apparently still in progress.

At Chiloe there has been a gradual elevation of about four feet in four years. These, together with numerous other examples, testify to the gradual but, as compared with other parts of the globe, exceedingly rapid rise of the ground upon the western shores of South America.[1] The most important point to be noticed is that this district of rapid elevation is one of the most earthquake-shaken regions of the world. And further, judging from Darwin's remarks, in those portions of it where the movements have been the most extensive, and at the same time probably the most rapid, the seismic disturbances appear to have been the most noticeable.

Similar remarks may be applied to Japan, it being in those districts where evidences of recent elevation are abundant that earthquakes are numerous. Thus, in the bay of Yedo, where we have borings of lithodomi in the tufaceous cliffs ten feet above high-water mark, which,

[1] Darwin, *Geological Observations*, p. 275 *et seq.*

inasmuch as the rock in which they are found is soft and easily weathered, indicate an exceedingly rapid elevation, earthquakes are of common occurrence.

From the evidences of elevation which we have upon the South American coast, Japan, and in other countries, it appears that these movements are intermittent, there being periods of rest, when sea cliffs are denuded, and perhaps even periods of subsidence. There is also evidence to show that, although these movements have been gradual from time to time, they have been aided by starts occasioned by earthquakes.

As to whether earthquakes are more numerous during periods of elevation, or of subsidence, or during the intermediate periods of rest, we have no evidence.

Sudden displacements which occasionally accompany earthquakes might, it was said, sometimes be regarded as the *cause* of an earthquake, and sometimes as the *effect*.

The slow elevations here referred to may be looked upon as being one of the more important factors in the production of earthquakes. By various causes the rocky coast is bent until, having reached the limit of its elasticity, it snaps, and, in flying back like a broken spring, causes the jars and tremors of an earthquake.

If this is the case, then the number of earthquakes felt in a district which is being elevated may possibly be a function of the rate of elevation.

APPENDIX.

LIST OF THE PRINCIPAL BOOKS, PAPERS, PERIODICALS, WHICH
ARE REFERRED TO IN THE PRECEDING PAGES.

*For a more complete bibliography of earthquakes refer to Mallet's
catalogue of works given in his report to the British Association in
1858.*

A True and Particular Relation of the Dreadful Earthquake which
   happened at Lima, &c. (1746).   1768.

Abbot, Gen. H. L.   On the Velocity of Transmission of Earth Waves.
   Am. Jour. Sci. XV., March 1878.

— Shock of the Explosion at Hallet's Point, Nov. 14, 1876.   *Batta-
   lion Press.*

Alexander, Prof. T.   See *Trans. Seis. Soc. of Japan.*

American Journal of Science.

Annali del reale osservatorio meteorologico Vesuviano.

Annual Register, The.

Anonymous, A Chronological and Historical Account of the most
   Memorable Earthquakes in the World, &c.   1750.

— A Vindication of the Bishop of London's Letter occasioned by
   the Late Earthquake.   1750.

— Phenomena of the Great Earthquake of Nov. 1, 1755.

— Serious Thoughts occasioned by the Late Earthquake at Lisbon.
   1755.

Asiatic Society of Japan, Transactions of.

Ayrton, Prof. W. E.   *See* Perry, J.

Bárceno, M. Estudio del Terremoto (May 17, 1879) Mexico.   1879.

Beke, Dr. C. T.   Mount Sinai a Volcano.

Bissett, Rev. J. A Sermon (on account of the Earthquake at Lisbon, Nov. 1, 1755). 1757.

Bittner, A. Beiträge zur Kenntniss des Erdbebens von Belluno vom 29. Juni 1873.

— Sitzungsb. der K. Akad. d. Wissensch., lxix. II. Abth., 1874.

Bollettino del Vulcanismo Italiano.

Boué, Dr. A. Ueber das Erdbeben welches Mittel-Albanien im October d. J. so schrecklich getroffen hat. *Die K. Akad. d. Wissenschaften*, Nov. 1851.

— Parallele der Erdbeben, des Nordlichtes und des Erdmagnetismus.

— Ueber die Nothwendigkeit die Erdbeben und vulcanischen Erscheinungen genauer als bis jetzt beobachten zu lassen. *Die K. Akad. d. Wissenschaften*, 1851 and 1857.

Bouguer, M. Of the Volcanoes and Earthquakes in Peru. British Association, Reports of.

Brunton, R. H. Constructive Art in Japan. *Trans. Asiatic Soc. of Japan*, II. and III., Pt. 2.

Bryce, J. Report to British Association, 1841.

Buffour, M. The Natural History of Earthquakes and Volcanoes.

C. H. A Physical Discussion of Earthquakes, &c. 1693.

Canterbury, Thomas, Lord Archbishop of. The Theory and History of Earthquakes.

Casariego, E. A. See *Trans. Seis. Soc. of Japan*.

Cawley, G. Some Remarks on Construction in Brick and Wood, &c. *Trans. Asiatic Soc. of Japan*, VI. Plate ii.

Chaplin, Prof. W. S. An Examination of the Earthquakes recorded at the Meteorological Observatory, Tokio. *Trans. Asiatic Soc. of Japan*, VI. Part ii.

Comptes Rendus.

Credner, H. Das Dippoldiswalder Erdbeben vom Oktober 1877.

— Zeitschr. f. d. Naturwiss. f. Sachsen u. Thüringen.

— Das Vogtländisch-erzgebirgische Erdbeben, 23. Nov. 1875.

— Zeitschr. f. d. gesammt. Naturwissenschaften, xlviii., Oktober.

Dan, T. See *Trans. Seis. Soc. of Japan*.

Darwin, Charles. Researches on Geology and Natural History.

— Geological Observations.

Darwin, G. H. Reports on Lunar Disturbance of Gravity to British Association, 1881. 1882.

Diffenbach, F. Plutonismus und Vulkanismus in der Periode von 1868–1872, und ihre Beziehungen zu den Erdbeben im Rheingebiet.

Doelter, C. von. Ueber die Eruptivgebilde von Fleims, nebst einigen Bemerkungen über den Bau älterer Vulcane.

— lxxiv. Band d. Sitzungsb. d. K. Akad. d. Wissensch., I. Abth., Dec. Heft, Jahrg. 1876.

Doolittle, Rev. T. Earthquakes Explained and Practically Improved, &c. 1693.

Doyle, P. See Trans. Seis. Soc. of Japan.

Emerson, Prof., B.A. Review of Von Seebachs' Earthquake of March 6, 1872. Am. Jour. Sci., Series III.

Ewing, Prof. J. A. Earthquake Measurement. A memoir published by the Tokio University. 1883.

— See Trans. Seis. Soc. of Japan.

Falb, R. Gedanken und Studien über den Vulcanismus, &c. 1875.

— Grundzüge zu einer Theorie der Erdbeben und Vulkanausbrüche.

— Das Erdbeben von Belluno. 'Sirius,' Bd. VI., Heft ii.

Flamstead, J. A Letter concerning Earthquakes. 1693.

Forel, F. A. Les Tremblements de Terre (Suisse). Arch. des Sciences Physiques et Naturelles, VI. p. 461.

— Tremblement de Terre du 30 Décembre 1879.

Fuchs, Karl. Vulkane und Erdbeben.

— Die Vulkanischen Erscheinungen der Erde.

Garcia, J. C. See Trans. Seis. Soc. of Japan.

Geinitz, Dr. E. Das Erdbeben von Iquique am 9. Mai 1877, &c. Die K. Leop.-Carol.-Deutschen Akademie der Naturforscher, Band xl., Nr. 9.

Gentleman's Magazine, The.

Geographical Society, Proceedings of.

Geological Society, Proceedings of.

Girard, Dr. H. Ueber Erdbeben und Vulkane. 1845.

Gray, T. See Trans. Seis. Soc. of Japan.

— On Instruments for Measuring and Recording Earthquake Motions. Phil. Mag. Sept. 1881.

— On Recent Earthquake Investigation. The Chrysanthemum, 1881.

Guiscardi, Prof. G. Notizie del Vesuvio. 1857.

— Il terremoto di Casamicciola del 4 Marzo. 1881.

Hales, S., D.D., F.R.S.   Some Considerations on the Causes of Earth-
    quakes.   1750.
Hamilton, Sir W.   Observations on Mount Vesuvius, Mount Etna,
    &c.   1774.
Hattori, I.   Destructive Earthquakes in Japan.   *Trans. Asiatic Soc.
    of Japan*, V. Plate i.
Heim, Prof. A.   Les Tremblements de Terre et leur Etude Scientifique.
    1880.
— Prof. A.   Die Schweizerischen Erdbeben in 1881–1882.

Hoeffer, Prof. H.   Die Erdbeben Kärntens und deren Stosslinien.
    Die Kais. Akademie d. Wissenschaften, Band xlii.
Höfer, Prof. H.   Das Erdbeben von Belluno, am 29. Juni 1873.
    Sitzungsb. der K. Akad. d. Wissensch., I. Abth., Band lxxiv.
Hoff, K. E. A. von.   Geschichte der durch Ueberlieferung nachgewie-
    senen natürlichen Veränderungen der Erdoberfläche.   1822.
Hooke, R., M.D., F.R.S.   Discourses concerning Earthquakes.
Hopkins, William.   Report to the British Association on the Geolo-
    gical Theories of Elevation and Earthquakes.   1847.
Horton, Rev. Mr.   An Account of the Earthquake which happened
    at Leghorn in Italy (Jan. 1742).   1750.
Humboldt, Alexander von.   Cosmos.
— Travels.

Jeitteles, L. A.   Bericht über das Erdbeben am 15. Januar 1858.
— Sitzungsberichte der mathem.-naturw. Classe d. K. Akad. d.
    Wissensch., xxxv. S. 511.
Judd, J. W., Prof.   Volcanoes, What they Are and What they Teach.

Knipping, E.   Verzeichniss von Erdbeben wahrgenommen in Tokio,
    &c.   *Mitt. d. Deutsch. Gesellsch. für Natur- und Völkerkunde
    Ostasiens*, Heft 14.
— See *Trans. Seis. Soc. of Japan.*

Lasaulx, A. von.   Das Erdbeben von Herzogenrath am 22. October
    1873.
Lemery, M.   A Physico-Chemical Explanation of Subterranean Fires,
    Earthquakes, &c.
Lescasse, M. J.   Etude sur les Constructions Japonaises, &c.   *Mé-
    moires de la Société des Ingénieurs Civils.*
Lister, M., M.D., F.R.S.   Of the Nature of Earthquakes.
Little, Rev. J.   Conjectures on the Physical Causes of Earthquakes
    and Volcanoes.   1820.

Mallet, R. The Neapolitan Earthquake, Vol. II. *Reports to the British Association*, 1850, 1851, 1852, 1854, 1858, 1861.
— Secondary Effects of the Earthquake of Cachar. *Proc. Geolog. Soc.*, 1872.
— Dynamics of Earthquakes. *Trans. Royal Irish Acad.* 1846.
Michell, Rev. J. Conjectures Concerning the Cause and Observations upon the Phenomena of Earthquakes. 1760.
Milne, David. Reports to British Association, 1841, 1843, 1844.
Milne, John. See *Trans. Seis. Soc. of Japan*.
— On Seismic Experiments (with T. Gray, B.Sc., F.R.S.E.) *Trans. Royal Soc.* 1882.
— On Seismic Experiments (with T. Gray, B.Sc., F.R.S.E.) *Proc. Royal Soc.* No. 217, 1881.
— Earthquake Observations and Experiments in Japan (with T. Gray, B.Sc., F.R.S.E.) *Phil. Mag.*, Nov. 1881.
— On the Elasticity and Strength Constants of certain Rocks (with T. Gray, B.Sc., F.R.S.E.) *Jour. Geolog. Soc.*, 1882.
— A Visit to the Volcano of Oshima. *Geolog. Mag.*, Dec. 2, Vol. IV., pp. 193–197, 255.
— On the Form of Volcanoes. *Geolog. Mag.*, Dec. 2, Vol. V., and Dec. 2, Vol. VI.
— Note upon the Cooling of the Earth, &c. *Geolog. Mag.*, Dec. 2., Vol. VII., p. 99.
— Investigation of the Earthquake Phenomena of Japan. *Rep. Brit. Assoc.*, 1881 and 1882.
— A Large Crater. *Popular Science Review*.
— The Volcanoes of Japan (a series of Articles). *Japan Gazette*.
— Earthquake Literature of Japan (a series of Articles). *Japan Gazette*.
— The Earthquake of Dec. 23, 1880. *The Crysanthemum*, 1881.
— Earthquake Motion. *The Crysanthemum*, 1882.
— Seismology in Japan. *Nature*, Oct. 1882.
— Earth Movements. *The Times*, Oct. 12, 1882.
Mohr, Dr. F. Geschichte der Erde. 1875.

Naturkundig Tijdschrift voor Nederlandsch Indie. 1875–1880.
Naumann, Dr. E. Ueber Erdbeben und Vulkanausbrüche in Japan. *Mitt. d. Deutsch. Gesellsch. für Natur- und Völkerkunde Ostasiens.* Heft 15.

354   APPENDIX.

Noggerath, Dr. J. Die Erdbeben vom 29. Juli 1846 im Rheingebiet, &c.
— Die Erdbeben im Vispthale (1855).
— Die Erdbeben im Rheingebiet in den Jahren 1868, 1869, 1870.
— Jahrgänge d. Verhandlungen d. Natur. Vereins für 1870. *Rhein-
    land u. Westphalen,* xxvii.

Oldham, Dr.   Secondary Effects of the Earthquake of Cachas. *Proc.
    Geolog. Soc.* 1872.
— Thermal Springs of India. *Memoirs of Geolog. Survey of India,*
    XIX. Plate 2.
— A Catalogue of Indian Earthquakes. *Memoirs of Geolog. Survey
    of India,* XIX. Plate 3.
— The Cachas Earthquake. *Memoirs of Geolog. Survey of India.*
    XIX. Plate 1.

Palmer, Col. H. S.   See *Trans. Seis. Soc. of Japan.*
Palmieri, Prof. L., e Scacchi, A.   Della Regione Volcanica del Monte
    Vulture, e del Tremuoto ivi avvenuto nel dì 14 Agosto 1851.
    1852.
— Annali del reale Osservatorio Meteorologico Vesuviano.
— Il Vesuvio, il Terremoto d' Isernia e l'eruzione sottomarina di
    Santorino. *R. Accad. d. Sci. Fis. e Mat. di Napoli,* iv. 1866.
— Sul recente Terremoto di Corleone. *R. Accad. d. Sci. Fis. e Mat.,*
    v. 1876.
— Il Terremoto di Scio del dì 4 Aprile. *R. Accad. d. Sci. Fis. e
    Mat. di Napoli.* v. 1881.
— Sul Terremoto di Casamicciola del 4 Marzo 1881. *R. Accad. d.
    Sci. Fis. e. Mat. di Napoli.* 1881.
Paul, Prof. H. M.   See *Trans. Seis. Soc. of Japan.*
Perrey, Prof. A.   Earthquake Catalogue and Memoirs.   (For list see
    Mallet's Report to British Association.   1858.)
— See *Trans. Seis. Soc. of Japan.*
Perry, J., and W. E. Ayrton.   On a Neglected Principle that may be
    Employed in Earthquake Measurement.
— See *Trans. Seis. Soc. of Japan.*
Philosophical Magazine.
Pickering, Rev. R.   An Address to those who have either retired
    or intend to leave Town under the Imaginary Apprehension of
    the Approaching Shock of another Earthquake. 1750.

Ray, J., F.R.S.   A Summary of the Causes of the Alterations which
    have happened to the Face of the Earth.

APPENDIX. 355

Rockstroh, E. Informe de la Comision Científica del Instituto Nacional de Guatemala, nombrada por el Sr. Ministro de Instruccion Pública para el Estudio de los Fenómenos Volcánicos en el Lago de Tlopango. 1880.

Rockwood, Prof. C. G. Notes on Earthquakes. Annually in the *Am. Jour. Sci.*

— Japanese Seismology. *Am. Jour. Sci.*, XXII. Dec. 1881.

Romaine, W. A Discourse occasioned by the Late Earthquake. 1755.

Rossi, Prof. M. S. di. Intorno all' odierna fase dei Terremoti in Italia, e segnatamente sul Terremoto in Casamicciola del 4 Marzo 1881. *Società Geografica Italiana.* 1881.

— La Meteorologia Endogena, 2 vols.

Royal Society, Transactions of.

Scacchi, A. *See* Palmieri.

Schmidt, Dr. J. F. Untersuchungen über das Erdbeben am 15. Januar 1858.

— Studien über Erdbeben. 1879.

— Die Eruption des Vesuv (1855). 1856.

Scrope, G. P. Volcanoes.

Seebach. Das mittle Deutsche Erdbeben (1872). *Mitt. der K.K. geograph. Gesellsch.*, II. Jahrg., 2. Heft, 1873.

Serpieri, Prof. A. C. S. Nuove Osservazioni sul Terremoto avvenuto in Italia il 12 Marzo 1873. *Istituto Lombardo.* 1873.

— Il Terremoto di Rimini della notte 17–18 Marzo 1875.

— Documenti nuove e Riflessioni sul Terremoto della notte 17–18 Marzo 1875. *Meteorologia Italiana*, iv. 1875.

— Determinazione delle fasi e delle leggi del grande Terremoto avvenuto in Italia nella notte 17–18 Marzo 1875. *Istituto Lombardo.* 1875.

— Dell' influenza del Lume Solare sui Terremoti. *Istituto Lombardo.* 1882.

Sherlock, T., D.D. (Lord Bishop of London). A Letter on the occasion of the late Earthquakes. 1750.

Shower, Rev. J., D.D. Practical Reflections on the Earthquakes that have happened in Europe and America, &c. 1750.

Stübel, A. (see Reiss, W.)

Stukeley, Rev. W., M.D., F.R.S. The Philosophy of Earthquakes, Natural and Religious, &c. Plates 1, 2, and 3. 1756.

Sturmius, J. C. A Methodical Account of Earthquakes.

A A 2

Suess, E. Die Erdbeben Niederösterreiches. *Die Kais. Akademie der Wissenschaften*, Bd. xxxiii.
— Die Erdbeben des südlichen Italiens. *Die Kais. Akademie der Wissenschaften*, Bd. xxxiv.

Volger, Dr. G. H. Untersuchungen über das Phänomen der Erdbeben. 1857.

Wagener, Dr. G. Bemerkungen über Erdbebenmesser und Vor-schläge zu einem neuen Instrumente dieser Art. *Mitt. d. Deutsch. Gesellsch. für Natur- und Völkerkunde Ostasiens*, Heft 15.
— See *Trans. Seis. Soc. of Japan*.

Winchilsea, The Earl of. A True and Exact Relation of the late Prodigious Earthquake and Eruption of Mount Etna. 1669.

Woodward, J., M.D., F.R.S. Earthquake caused by some Accidental Obstruction of a Continual Subterranean Heat.

SEISMOLOGICAL SOCIETY OF JAPAN.

The following are a list of the papers published by this Society:—

VOL. I.

Milne, J. Seismic Science in Japan. 35 pages.

Ewing, J. A. New Form of Pendulum Seismograph. 6 pages, 3 plates.

Gray, T. Seismometer and Torsion Pendulum Seismograph. 8 pages, 2 plates.

Mendenhall, T. C. Acceleration of Gravity at Tokio (abstract). 2 pages.

Wagener, G., and E. Knipping. New Seismometer and Obervations with same. 18 pages, 1 plate.

Milne, J. Earthquake in Japan of Feb. 22, 1880. 116 pages, 5 plates, 8 woodcuts.

VOL. II.

Milne, J. Recent Earthquakes of Yeddo, Effects on Buildings, &c. 38 pages, 2 plates, and many tables.

Mendenhall, T. C. Gravity on Summit of Fujiyama (abstract). 2 pages.

Paul, H. M. Earth Vibrations from Railroad Trains (abstract). 4 pages.

Ewing, J. A. Astatic Horizontal Lever Seismograph (abstract). 5 pages, 1 plate.

Milne, J. Peruvian Earthquake of May, 9, 1877. 47 pages, 2 plates, tables. Constitution, Rules, Officers and Members of the Society, Dec., 1881.

VOL. III.

Gray, T. Steady Points for Earthquake Measurements. 11 pages, 3 plates.

Milne, J. Experiments in Observational Seismology. 53 pages, 1 plate, tables.

— The Great Earthquakes of Japan. 38 pages, 1 plate, many tables.

Perry, J. Theory of a Rocking Column. 4 pages.

Knipping, E. Earthquake of July 25, 1880, with Dr. Wagener's Seismometer. 4 pages.

Ewing, J. A. Earthquake Observation at three or more Stations, &c. 4 pages.

— Records of three recent Earthquakes. 6 pages, 3 plates.

— Earthquake of March 8, 1881. 8 pages, 1 plate.

Milne, J. Horizontal and Vertical Motion in Earthquake of March 8, 1881. 8 pages, 3 plates.

Gray, T. Seismograph for Registering Vertical Motion. 3 pages, 1 plate.

Ewing, J. A. Seismometer for Vertical Motion. 3 pages, 1 plate.

Gray, T. Seismograph for Large Motions. 2 pages.

— Compensating a Pendulum to make it Astatic. 3 pages.

Palmer, H. S. Note on Earth Vibrations. 3 pages.

Kuwabara, M. The Hot Springs of Atami. 2 pages.

VOL. IV.

Milne, J. Distribution of Seismic Activity in Japan. 30 pages, 1 plate.

Wada, T. Notes on Fujiyama. 7 pages.

Casariego, E. Abella y. Earthquakes of Nueva Vizcaya in 1881. 23 pages, 2 maps.

Milne, J. Utilisation of Earth's Internal Heat. 12 pages.

Ewing, J. A. Earthquake of March 11, 1882. 5 pages.

Doyle, P. Note on an Indian Earthquake. 6 pages.

Milne, J. Systematic Observation of Earthquakes. 31 pages, 5 plates.

VOL. V.

Naumann, Dr. E. Notes on Secular Changes of Magnetic Declination in Japan. p. 1–18.

Casariego, Don E. Abella y. Monografía Geológica del Volcan de Albay ó El Máyon. p. 19–43.

INDEX.

PRINTED BY
SPOTTISWOODE AND CO., NEW-STREET SQUARE
LONDON

Printed in the United States
By Bookmasters